Nietzsche's Dancers

Nietzsche's Dancers

Isadora Duncan, Martha Graham, and the Revaluation of Christian Values

Kimerer L. LaMothe

palgrave
macmillan

NIETZSCHE'S DANCERS
Copyright © Kimerer L. LaMothe, 2006.

First published in hardcover in 2006 by
PALGRAVE MACMILLAN®
in the United States—a division of St. Martin's Press LLC,
175 Fifth Avenue, New York, NY 10010.

Where this book is distributed in the UK, Europe and the rest of the world,
this is by Palgrave Macmillan, a division of Macmillan Publishers Limited,
registered in England, company number 785998, of Houndmills,
Basingstoke, Hampshire RG21 6XS.

Palgrave Macmillan is the global academic imprint of the above companies
and has companies and representatives throughout the world.

Palgrave® and Macmillan® are registered trademarks in the United States,
the United Kingdom, Europe and other countries.

ISBN: 978–0–230–33844–9

Library of Congress Cataloging-in-Publication Data

LaMothe, Kimerer L.
    Nietzsche's dancers : Isadora Duncan, Martha Graham, and the
    revaluation of Christian values / by Kimerer L. LaMothe.
        p. cm.
    Includes bibliographical references and index.
    ISBN 1–4039–6825–X
    1. Dance—Philosophy. 2. Nietzsche, Friedrich Wilhelm,
    1844–1900—Influence. 3. Dance—Religious aspects—Christianity.
    I. Title.
GV1588.L315 2006
792.8_01—dc22   2005050221

A catalogue record of the book is available from the British Library.

Design by Newgen Imaging Systems (P) Ltd., Chennai, India.

First PALGRAVE MACMILLAN paperback edition: October 2011
Transferred to Digital Printing in 2011

Contents

Abbreviations

Friedrich Nietzsche

AC   "The Antichrist" in *The Portable Nietzsche*
BGE   *Beyond Good and Evil*
BT   *The Birth of Tragedy*
CW   *The Case of Wagner*
EH   *Ecce Homo*
GM   *On the Genealogy of Morals*
GS   *The Gay Science*
HH   *Human All Too Human*
K   *Kritische Studienausgabe*
OTL   "On Truth and Lying in their Extra-moral Sense"
TI   "Twilight of Idols" in *The Portable Nietzsche*
WP   *The Will to Power*
Z   "Thus Spoke Zarathustra" in *The Portable Nietzsche*

Isadora Duncan

AD   *Art of the Dance*
IS   *Isadora Speaks*
ML   *My Life*

Martha Graham

B   *Blood Memory*
G#   Essays by Graham
N   *The Notebooks of Martha Graham*

Preface

How does it all begin?
I suppose it never begins, it just continues—
Life—
generations
Dancing—N 302
And one . . . —B 276

The question that impels the writing of this book concerns bodies, our human bodies. I want to know how we, as human beings in the twenty-first century, should live our relationship to the bodily dimensions of our being. Of what value are our bodies?

Friedrich Nietzsche would caution me here. There is a limit to what I can ask and expect in this regard. To separate the "I" from "the body" in asking this question is to posit a false distinction. As he writes in *Thus Spoke Zarathustra*, the "I" is nothing but the body: "the body and its great reason: that does not say 'I,' but does 'I' " (K4, 40; Z 146). To believe in the body as a thing, as some material entity that exists apart from my self, is to make an error. It demonstrates a faith that words—linguistic symbols—correspond to what exists. Rather, Nietzsche would remind me, there is no moment when "I" am not a body in action, no place outside my embodiment where "I" can stand, as all-seeing master, willfully commanding its obedience. My questioning "I" represents nothing more or less than my particular experience of embodiment.

So reminded, "I" clarify. What I seek is not an ideal of thought or action to which people must conform their bodies. Rather, I am interested in learning how to participate in the ongoing process of bodily becoming that my "I" expresses. I/body am always already a process of turning bodily sensations into metaphors, images, and ideas, including my idea of "I." I/body is an ongoing dynamic of reflexivity and self-creation. What I/body sense and perceive and transform is not *mine* in any strict sense; it represents all the training and experience mediated to me by family, culture, history, and tradition. Nevertheless, how I/body invest energy in these processes of transformation is what guides me in becoming who I am. In short, I seek to understand and critique the range of forces—social and psychological, cultural and physiological—shaping how "I" perceive and conceive the experience of being "body."

What role does or should our bodily being play in the process of becoming who we are? What if the conceptions of body to which we aspire are diminishing our bodily health—our ability to sustain those very conceptions as true? What if our sense of I is inhibiting our ability to unfold our energetic potential? Of what value is it to be a body—or an I?

*   *   *

Nietzsche's Dancers enlists materials left behind by a dancing philosopher and two philosopher dancers to help pursue these questions: Friedrich Nietzsche, Isadora Duncan, and Martha Graham. While this project may not unearth new facts about the lives of these figures, it discloses new possibilities of interpretation by calling attention to dimensions of their work often overlooked by scholars in the fields of philosophy, religious studies, and dance studies. All three of these individuals were deeply concerned about human embodiment as valued, or devalued, by Christian theology, morality, and its offspring, modern science. All three challenged the faith in verbal practices that justifies the marginal status of arts in scholarship. Moreover, all three demonstrated a persistent commitment to employing *dance*—as image, practice, and art—not only to critique the Christian attitudes toward bodily being they observed, but to develop alternative evaluations of bodily being in relation to what they understood as "religion." For all three figures, as this book demonstrates, the well-being of humanity depended upon the mutual critique and evolution of the forms by which dance and religion have been practiced, studied, and known in Christian cultures. These three figures insisted, though differently, that we in the contemporary world must overcome a longstanding hostility toward dance propounded by Christian religion, a hostility that is ensconced in our cultural institutions and sustained in our personal, most intimate senses of ourselves as minds, souls, or spirits dwelling in bodies.

The fact that these three individuals' claims for religion and dance are still overlooked bears witness to the radical challenge they pose. As the Introduction points out, most commentators insist that Nietzsche could not have been speaking literally when he describes free spirits as dancing. Nor could Duncan have meant that her dancing would actually catalyze a "renaissance of religion," or Graham that her dancers were really "athletes of God." In each case, commentators conclude, the person was speaking poetically, for rhetorical effect. Yet these responses to their work prove that Nietzsche, Duncan, and Graham were right about the depth at which we

believe that what words *can* represent—namely, a difference between "I"
and "body"—is therefore real and beneficial. Nietzsche, Duncan, and
Graham sought to elevate dance alongside verbal practices as an alternate
medium for generating ideas of and about religion—ideas capable of
affirming the dynamic, creative becoming of our bodily being, and its I.

* * *

Although *Nietzsche's Dancers* begins with Nietzsche's texts, this project as a
whole did not. While I first read Nietzsche as a junior in college, it was not
until graduate school that I realized the extent to which Nietzsche's writing
abounds with dance imagery. By that time, I had discovered my passion for
modern dance: I knew I was a dancer, and that I would never not be. I was
choreographing and performing professionally and training regularly while
completing doctoral work in Christian theology. Moving between these
fields of dance and religious studies, I was frustrated with the absence of
dance references in modern western philosophy and theology. Coming
upon Nietzsche's texts again at this time, I welled with relief. For me, his
dance metaphors were not merely linguistic devices—they represented an
acknowledgement of the experiences that would cause those images to have
meaning at all. Reading Nietzsche gave me permission to value my experiences of dancing as relevant to my scholarship in philosophy and theology
of the modern West.

Later, while reading about the modern dancers, I realized that I was not
the only dancer who had appreciated Nietzsche. The founders of the
modern dance I loved to do had also been similarly inspired. In fact, it was
Duncan and Graham's persistent celebration of Nietzsche's work that
encouraged me to think more extensively about what his images of dance
might contribute to the philosophy of religion. Yet Duncan and Graham
did not respond to Nietzsche in words alone. They made dances. Not only
that—they evolved new techniques of dance practice and performance.
They pioneered an aesthetics that was later recognized as "modern" dance
(a title that neither Duncan nor Graham initially used). Thus, my interest
in discerning what their work would help me see in Nietzsche impelled me
to develop conceptual resources for studying dance as a medium for generating ideas about religion.

So began the oscillating movement between Nietzsche and the modern
dancers that came to constitute the narrative of this book. My reading of
Nietzsche's texts represents my experiences as a modern dancer, my intellectual and kinetic studies of dance technique and choreography by Duncan
and Graham, and my conviction that the insights to which these experiences lead me are crucial to my work in religious studies. Conversely, what

I find in the work of Duncan and Graham derives from my conviction that the significance of their innovations cannot be calculated without reference to the ways in which they contribute to twentieth, and twenty-first, century conversations concerning the life and value of religion, Christianity in particular. Thus, in every sentence on every page, my argument moves back and forth among Nietzsche, Duncan, and Graham, even when that movement is not explicit. At the same time, my intent is not to privilege writing over dancing or dancing over writing; nor is it to assert the identity of Nietzsche's philosophizing and Duncan and Graham's dancing. I rather juxtapose these figures so as to highlight the differences among them along points of mutual concern—namely, the revaluation of Christian attitudes toward bodily being.

*   *   *

That these questions concern me now is a sign of where I/body find myself as a dancer and scholar in religious studies. As many have noted, one mark of contemporary scholarship is precisely that "the body" has become a problem in a new way. The body serves as both subject and object while the wedge driving apart these two manifestations grows larger daily.

On the one hand, it is now taken for granted that all human knowledge is mediated to us through five senses, our physical insertion in the manifold of the given. Our science gathers information through these senses, and develops mechanical means for extending and refining their range. Knowledge comes through careful analysis of the sensory data.

On the other hand, the body itself appears as an object of study, and one of the most interesting. It is the materiality that we are most intent to deconstruct, decipher, cure, or purify. We treat our bodies as objects when we exercise them, regulate them, decorate them, and surgically modify them in the name of science, beauty, or health.

The contemporary bind, then, is this: scholars in all fields invest their time in *thinking about bodies* via data provided to them through various logical and technological extensions of human sense, books included; and in doing so they practice *ignoring the experience of their own bodies* as a precondition for the objective status of their scholarship.

This contradiction holds for people in most walks of contemporary life: we are increasingly unable to experience our own bodies as sites and sources of knowledge. We are unable to experience our selves as ongoing processes of bodily becoming, or to experience our bodily becoming as an ongoing process of our selves. Moreover we cultivate this ignorance as the condition that will enable us to make our bodies into the images we desire for them.

This predicament is nowhere more evident, perhaps, than in the study and practice of Christian religion. Christian theology represents a pinnacle of linguistic abstraction: "God" appears as the word for an ultimate ideal, the full presence of Being, an extension to infinity beyond all materiality or all values. Not surprisingly, the dominant (though not the only) model for spirituality across Christian denominations is linguistic: a Christian must read, study, hear words in order to welcome the Word, the incarnation of God in Jesus, into her heart. Through such verbal practices, the Word will take root and grow into new forms of compassion and love, thought, feeling, and action. For Christians who abide by this model, whether conservative or liberal, Catholic or Protestant, there is little sense that cultivating a relationship with God might entail cultivating a physical consciousness of one's own body as a rhythm of its own *becoming*.

A similar model continues to prevail in religious studies where scholars work to read symbolic phenomena (whether texts, rituals, or artifacts) and translate them into rational forms that can be written or spoken in scholarly forums. When applied to dance, such approaches fail to recognize bodily movement as making a distinct contribution to religious life, except where the movement appears to function as words do (LaMothe 2004; 2005c). While there is vociferous disagreement in such contexts over how and what words mean, and over who is authorized to decide, the effect of these debates is to solidify participants' conviction that the meaning of words is the issue.

*   *   *

While many claim to discern the contribution Nietzsche makes to philosophy, theology, and history in general, his value for me lies in his persistent resurrection of bodily becoming as a process through whose rhythms all forms of human culture, including philosophy and religion, come into being. Like Duncan and Graham who follow him in this (if not every) way, Nietzsche interrogates the hegemony of written words over the western theological imagination. He calls into question our faith in words as conduits for truth; he demonstrates in painstaking and often repetitive detail how this faith finds expression in a morality and science that encourage hostility toward life, ignorance of bodily becoming, and unproductive violence toward ourselves and others. Faith in practices of reading and writing perpetuates a dualistic way of thinking about our selves that occludes creative participation in our own becoming. When we sever our relation to our embodiment, thinking ourselves free, we generate fear and hatred of our uncontrollable bodies. We feel justified in violating others as we violate ourselves. Such faith, Nietzsche contends, also prevents us from exercising the muscles we must develop in order to have the strength to affirm life in all its suffering and joy. In the effort to seek ease and comfort, we tear ourselves and our worlds to pieces. Thus, in maintaining faith in practices of reading and writing, we are losing the capacity to make choices for ourselves and others that honor the bodies we presume to value, whether we conceive of them as creatures of God or subjects and objects of scientific study. In Zarathustra's words, *we do not love ourselves enough* (K4, 77; Z 173). We do not dance.

It was here, perhaps, that Nietzsche was most helpful for Duncan and Graham in their respective efforts to make new and significant dance. Nietzsche envisioned an alternative mode of valuing embodiment, one he fleshes out with images of dance. Zarathustra incarnates this alternative valuation: he loves human beings because they are bodies becoming, and he is a dancer. In reading Nietzsche, Duncan and Graham found philosophical justification for developing their practice and performance of dancing alongside writing as making critical, creative contributions to cultural discussions concerning Christian values. In bringing these three individuals together within the scope of this book, I/body trace the movements of their visions in the hope of inspiring others to continue their work.

* * *

A project that spans disciplines as this one does owes its life to many people and institutions. It is impossible to say where it begins or ends. Its roots reach deep into my past as a child who loved to move and felt horribly confined by the hard pews in the Protestant churches she attended. Its branches reach far toward my ideals for what is possible in the future.

In the academic world I have benefited greatly from the support of teachers, mentors, and colleagues who have enabled and encouraged me to forge ahead in developing new approaches to scholarship in religious studies. A list, inevitably incomplete, must include: Mark C. Taylor, H. Gans Little, Gordon Kaufman, Lawrence Sullivan, Diana Eck, Margaret Miles, Sharon Welch, Francis Fiorenza, John Carman, Stanley Cavell, Diana Apostolos-Cappadona, Rita Nakashima Brock, Karen King, and Robert Orsi. Tyler Roberts graciously offered his insightful comments on earlier versions of the chapters on Nietzsche. In the world of dance, I am forever grateful for the training and inspiration I have received from Sharron Rose, Marcus Schulkind, Christine Dakin, Maher Benham, Denise Vale, Pearl Lang, Sylvia Gold, Elizabeth Bergmann, Deborah Foster, and Claire Mallardi.

Several institutions provided valuable support at various stages of the project, including the Radcliffe Institute, where I was one of the last

"Bunting" Fellows in 2000–2001; and the Religion and Arts Initiative at the Center for the Study of World Religions, where I received a Fortieth Anniversary Fellowship in 2003–2004. These communities provided enabling contexts for the initial and middle stages of this project, respectively. I am grateful to Gail Hamner of the "Feminist Theory and Religious Reflection" section of American Academy of Religion; to Brad Epps and Mari Ruti of the Harvard Humanities Center, and to the Women in Philosophy section of the American Philosophical Association for inviting me to present earlier versions of the ideas that appear here.

Family and friends form the crucial network of support and love that infuses this work with a sense of meaning. My parents, Jack and Cynthia LaMothe, siblings John LaMothe and Barrett LaMothe Ladd, as well as Suzanne Evenson and Janet Cree, have each, at crucial moments, provided this project with a needed push. Kathleen Skerrett, Colby Devitt, Emily Hudson, Deborah Abel, Mieke van der Wansem, Julia Foulkes, and Courtney Bickel Lamberth have engaged with me in conversations that have helped clarify my ideas and analysis. Along the way, I have depended on the excellent care provided by Michelle Hamblin for my youngest daughter, Kyra.

Above all I am hopelessly indebted to the members of my immediate family for supporting every step of this project. Jordan (who is pleased to share a birthday with Nietzsche), Jessica (who is writing her own books), and Kyra (who loves to dance) have welcomed this project into our family as one of their own. Their excitement at its arrival heightens mine. Finally, I am deeply grateful to Geoffrey, partner in life, whose generosity, tolerance, flexibility, and love ever inspire me to embrace as worthy the creative possibilities of bodily becoming.

Introduction: Reading Nietzsche's Images of Dance

We are something different from scholars, although it is unavoidable for us to be also, among other things, scholarly. We have different needs, grow differently, and also have a different digestion: we need more, we also need less. How much a spirit needs for its nourishment, for this there is no formula; but if its taste is for independence, for quick coming and going, for roaming . . . it is better for such a spirit to live in freedom with little to eat than unfree and stuffed. It is not fat but the greatest possible suppleness and strength that a good dancer desires from his nourishment—and I would not know what the spirit of a philosopher might wish more to be than a good dancer. For the dance is his ideal, also his art, and finally also his only piety [Frömmigkeit], his "service of God" [Gottesdienst].

—K3, 635; GS #381, 345–6

On the pages of Nietzsche's texts, multitudes dance. Dionysian revelers, satyrs of the tragic chorus, and Dionysus himself; medieval Christians, free spirits, inspiring muses, and Zarathustra; gods and goddesses, young girls, women, and higher men—all dance. So too do thoughts, words, pens, stars, and sometimes even philosophers. The dances performed are as diverse as the performers, differing in rhythm, style, context, and meaning. Nietzsche uses *Tanz* or "dance" to describe the activity that occurs during Greek festivals, satyr plays, and Attic tragedies; in the Bible and in Christian history; in social settings and crowds. Dance appears as art, religion, and recreation; as a discipline, a language of gestures, and an experience of rapturous intoxication. In nearly all of Nietzsche's books, from *Birth of Tragedy* to *Ecce Homo* and the posthumous *Will to Power*, images of dancers and dancing appear with remarkable consistency. Why? Why *dance*? What significance do these images carry in the context of Nietzsche's philosophy?

The task of interpreting these images is daunting and only a few have attempted it.[1] Even though some of the most frequently quoted phrases from Nietzsche's corpus contain references to dance, the significance of *dancing* to these passages is rarely addressed. The conversation quickly turns to the "spirit of gravity," to "light feet," or to the question of what constitutes life-affirming values. The dancing itself is left behind as irrelevant to the discussion it catalyzes. Even in the indices of translations, "dance" is rarely included.[2]

Yet close attention to the dance imagery across Nietzsche's work reveals a pattern of use that illuminates his philosophical project. More often than not, Nietzsche uses dance images strategically: they appear at crucial junctures in his arguments where he is articulating the relationship to Christian morality that he, as a philosopher and free spirit, seeks to embody. The relationship he seeks is one that is characterized by the term "revaluation" [*Umwertung*]. More often than not, Nietzsche's dancers are directly engaged in the process of *revaluing* Christian values. In their dancing, they are *overcoming* the Christian values alive in their bodies, and developing the energy and strength needed to *affirm life*, in all its suffering and joy, even when conceived as *recurring eternally*. "Dance" repeatedly appears to signal both the means and the fruit, the practice and the performance, of creating life-affirming values and learning to *love* life. What then can Nietzsche's dance imagery teach us about his project?

Two readers, who were also dancers can help answer this question. In the early twentieth century, Isadora Duncan and Martha Graham, read Nietzsche and took his call to dance seriously; they sought to create dances that were capable of realizing his vision of dance as a catalyst in revaluing religion. What they discovered in the process of developing their respective techniques of dance practice and performance provides resources for understanding why Nietzsche uses "dance" to represent a relationship to the body that revalues what he perceives as Christian hostility to bodily life.

Defusing Dance

The primary reasons why only a few in the field of Nietzsche scholarship have paid much attention to his dance images are coextensive with the history of the field. In philosophy and religious studies, an authority no less than Martin Heidegger set the terms within which Nietzsche studies continue to unfold. What Heidegger did would have pleased Elisabeth Förster-Nietzsche, Nietzsche's younger sister, sole survivor and executor until her death of the Nietzsche Archives. In Heidegger's two-volume *Nietzsche* (1961), derived from lectures and essays from 1936 to 1944, he crowned Nietzsche a great metaphysician (if the last), and based this claim on the collection of Nietzsche's unpublished notes that Elisabeth collected and edited after her brother's death under the title *Will to Power*.

In so doing Heidgegger initiated the debates over content and style that remain central to scholarly discussions over the relevance of Nietzsche's work. Nearly every commentary on Nietzsche situates itself in relation to questions posed by Heidegger's thesis and method: Is Nietzsche a philosopher or not? Is his project a metaphysics or an anti-metaphysics? Which writings of Nietzsche's—published and otherwise, literary or philosophical—represent the core of his project? Is he a nihilist, immoralist, atheist, relativist (whether epistemological, moral, or cultural) or not? Scholars carry out the terms of these questions in relation to Nietzsche's ethics, metaphysics, political theory, aesthetics, and philosophy of religion. Referring to "significant readings" of Nietzsche since Heidegger, Krell admits that "none of these writers can readily separate the names Nietzsche/Heidegger. None can pry apart this laminate. As though one of the crucial confrontations for thinkers today were what one might call heidegger's nietzsche, nietzsche's heidegger" (Heidegger 1991: xxvii). In establishing Nietzsche as a metaphysician, not surprisingly, Heidegger does not pay heed to Nietzsche's images of dance.

In the wake of Heidegger's reading, two currents of Nietzsche scholarship emerged that continue to dominate the field.[3] A first wave of commentators aligns with Heidegger to a greater and lesser extent in defending the *coherence* of Nietzsche's writing as a *philosophy* against charges of him as a nihilist, immoralist, or Nazi-enabler. Scholars in this stream (including Walter Kaufmann, Arthur Danto, Richard Schact, Maudemarie Clark, John Richardson, and others) often assert a different center than *Will to Power* or suggest a different periodization of writings around which they organize their presentation of Nietzsche's "philosophy" and their arguments about his conceptions of truth, knowledge, and value. A second stream of commentators (including Jacques Derrida, Sarah Kofman, Alexander Nehamas, Gilles Deleuze, Eric Blondel, and others) resist Heidegger's conclusions, calling attention to Nietzsche's styles and metaphors—the forms of his writing—as challenging the possibility of accomplishing metaphysical projects at all. There are also conversations that cross these currents, as scholars engage Nietzsche's writings in relation to philosophical issues or problems. Two crosscurrents relevant to this book include the strong tradition of commentary on Nietzsche and the "feminine" as well as more recent studies of Nietzsche as a religious thinker, or a thinker of religion.[4]

Despite the burgeoning health of the field, however, Nietzsche scholars have paid scant attention to his images of dance. For those who seek to discern philosophical principles and programs in Nietzsche's texts, dance images tend to appear as one more manifestation of Nietzsche's rhetorical flourish. When these scholars notice the images, they tend to interpret them as symbolizing Nietzsche's concepts without adding content to these concepts. Moreover, in the attempt to establish Nietzsche's coherence, scholars are apt to recognize as "coherent" what they have been trained to perceive. In so far as they have been trained in a tradition that, prior to Nietzsche (and in some places, even to this day), rarely considers "dance" as art, philosophy, or religion, most philosophers and theologians are not prepared to accept or even notice dance as relevant to the kind of questions they ask concerning truth, knowledge, and value.[5]

On the other hand, scholars who privilege Nietzsche styles and metaphors as keys to his philosophical import also downplay his dance imagery. While training and tradition are partly to blame, three brief examples elucidate additional dynamics at work in these approaches, and trace those dynamics to the terms of discussion set by Heidegger.

Kofman, Derrida, and Blondel

In the first brief chapter of *Nietzsche and Metaphor*, appearing in 1972, Sarah Kofman acknowledges how paradoxical it is "to make use of concepts in writing on a philosopher who privileges metaphor" (1993:1). She then launches into the second chapter by citing a passage from *Birth of Tragedy* that features *dance* as exercising the entire "symbolism of the body." However, in the theory of metaphor Kofman proceeds to distill from Nietzsche's texts, she does not engage his dance images. Kofman discusses architectural metaphors (beehives, towers, pyramids, spider webs), nakedness, and writing and reading. Moreover, in her discussion of these metaphors, she situates their meaning firmly within the realm of linguistic activity. These metaphors acquire meaning in the text as a function of their relationship with one another. After inaugurating the discussion of metaphor, dance disappears.

The metaphor of dance does eventually return—in the penultimate sentence of the book—where Kofman stages her conclusion: "Accepting to write, in the knowledge that the whole human race also writes in us, means not taking *ourselves* 'seriously' anymore, that is the wisdom of the *gay science* combined with laughter, lightness, and dance" (119). In this passage, Kofman herself uses dance as a metaphor but with much less carry over than Nietzsche. In her usage dancing appears as something extra or accidental, "combined with" the *gay science* of writing. It signifies a mental attitude: not taking oneself seriously. Dissipated in this usage is the tension between dancing and writing as different bodily activities, different "symbolisms," that Nietzsche introduces in *Birth* and develops in his later works—a tension designed to dislodge our concept and practice of *writing* as a foundation for our understanding of metaphor.

This dynamic—in which a scholar reads Nietzsche's metaphors as linguistic devices and then uses dance as a metaphor without commenting upon the fact—is the dominant mode among Nietzsche commentators for dealing with dance imagery. Scholars do not exactly ignore the images, but appropriate and use them in support of an interpretation of Nietzsche that denies the dance imagery any constitutive role in his project.

Following this dynamic in Derrida's work reveals its links to Heidegger. In the essay "Différance" from *Margins of Philosophy*, first published in 1972, Jacques Derrida makes the point that Nietzsche prefigures both Heidegger's critique of the metaphysics of presence and Derrida's own discovery of *différance* (1982:17). As Derrida writes: "*différance* is the name we might give to the 'active,' moving discord of different forces, and of differences of forces, that Nietzsche sets up against the entire system of metaphysical grammar, wherever this system governs culture, philosophy, and science" (18). Derrida continues that "This unnameable is the play which makes possible nominal effects of the relatively unitary and atomic structures that are called names" (26). The play of the unnameable finds expression in "nominal effects," words or names whose meaning is, in the end, undecidable. In *Spurs*, Derrida develops this point in his interpretation of Nietzsche's designation of truth as "a woman." Derrida concludes that Nietzsche uses contradictory remarks about "woman" to communicate the undecidability of truth, and thereby deflects all attempts—Heidegger's in particular—to prove him a metaphysician (1979:49–50).

Across Derrida's work he uses the metaphor of dance to represent this movement of *différance* and the consequent undecidability of truth. He uses the metaphor of dance to represent. As Derrida writes in "Différance," "There will be no unique name, even if it were the name of Being. And . . . we must *affirm* this, in the sense in which Nietzsche puts affirmation into play, in a certain laughter and a certain step of the dance" (1982:27).[6] In this passage, Derrida uses dance as a metaphor for a mental attitude capable of affirming the undecidability of truth, and does so in the context of slating metaphor itself as an effect of *writing*. While Derrida insists that he means for "writing" to include discursive and nondiscursive forms of communication or marking, his use of writing as an umbrella category serves to reinforce the privilege accorded in the West to the practice of discursive writing as a discipline for thinking.[7] Writing ranks high for Derrida because it breeds consciousness of its inability to secure truth. There is little reason to think that Derrida believes that Zarathustra means what he says when he urges the higher men to learn to *dance*.

The influence of Heidegger on Derrida's use of dance images is evident. Dance cannot appear in its own right as a bodily practice or as symbolic bodily movement because it does not respect the difference that Derrida wants to draw between Heidegger's reading of Nietzsche and his own. Derrida's thesis hinges on the (metaphorical) equation of *différance* with a concept of *writing* modeled on the act of using words. Thus, the act of excluding dance from consideration is part of what Derrida shares with Heidegger as the condition for opposing Heidegger's reading of Nietzsche. "Dance" can only appear as an (verbal) image after the hegemony of writing as a verbal practice over the philosophical imagination has been established.

As a final example, Eric Blondel, in *Nietzsche: The Body and Culture* (1986), attends to Nietzsche's bodily metaphors and interprets them as attempts to write the body, that is, attempts to write a text as a body and thus bring the body into philosophical discussion. His explicit intent is to resist those who textualize Nietzsche's writing, as I would argue both Heidegger and Derrida do (7–10). Nevertheless, Blondel does not include Nietzsche's dance imagery as relevant to Nietzsche's sense of what a body is, or how it is written. The effect of Blondel's reading, despite his intention, is to erase any sense of the body as having life or agency outside of the text, or at least, outside of the practices of reading and writing. Either a body appears in the text, as a text, as written by the text, or it is "mad" (31). The act of writing texts appears as "the signifying process" in and through which a "body" acquires significance (29).

In sum, for the few scholars who hazard remarks about Nietzsche's dance imagery, the dominant currents within Nietzsche scholarship encourage them to interpret those images as one figure among others whose meanings derive from the nexus of written images within which they appear. Dance most often appears as a sign of an ability to *think* in a particular way about values, knowledge, and truth.

Nietzsche himself seems to support such readings, as when he writes: "Learning to *think* . . . logic as a theory, as a practice, as a *craft* is beginning to die out . . . there is no longer the remotest recollection that thinking requires a technique, a teaching curriculum, a will to mastery—that thinking wants to be learned like dancing, *as* a kind of dancing. Who among Germans still knows from experience the delicate shudder which light feet in spiritual matters send into every muscle?" (K6,109; TI 512). A second look at this passage, however, argues against the conclusions most frequently drawn about Nietzsche's dance images. Nietzsche is not just saying that we should learn thinking "like" or "on the model" of dancing. He does say that, but he continues to make another statement. He urges readers to learn thinking *als eine Art Tanzen*—as a *kind of* dancing—that is, as a species or subset of dancing. Here thinking appears as a bodily discipline requiring a physical technique. The implication of this statement is that a person who wants to know how to think must know *how* to dance, else he will not know how to think as a *kind* of dancing.

Dancing Readers

Not all of Nietzsche's readers have read in the shadow of Heidegger. Early in the twentieth century, a handful of dancers—women and men—responded to Nietzsche's calls for readers with the long legs and stamina needed to jump from aphorism to aphorism. These readers interpreted Nietzsche's references to dance as relevant to what they were doing in their attempt to invent new forms of dance practice and performance. They embraced Nietzsche as a philosopher of dance, and found in his work resources for resisting the hostility to their work they experienced from representatives of Christian theology, morality, and tradition. Isadora Duncan, Ruth St. Denis, Mary Wigman, Ted Shawn, Doris Humphrey, Charles Weidman, and Martha Graham were some of those who read Nietzsche and described their dance-making, to a greater or lesser extent, as participating in his project of revaluing Christian values. As dance critic Deborah Jowitt notes, "Nietzsche was their lodestar" (1988:165–7).

Among these dancing readers, Duncan and Graham registered the deepest responses to his work. They expressed most forcefully and consistently the impact that reading Nietzsche had on both their visions for dance and their process of realizing those visions in the development of dance techniques and choreographies. For this reason, it is their stories that occupy the second and third parts of this book. Their work in dance provides a wealth of resources for interpreting Nietzsche's dance images, even as the connection to Nietzsche illuminates the philosophical significance of that work in dance.

Isadora Duncan (1877–1927) read Nietzsche in her early twenties.[8] By that time Duncan was already an international star, famous for the solo dances that she performed to live classical music in loose gauzy tunics, in front of a blue or green–gray curtain. She had rejected the aesthetics, training, and philosophy of classical ballet, and was on her way to developing what would become one of the first techniques of American modern dance. She was impelled in this process by a vision of a dance that would fulfill its potency as art by catalyzing a renaissance of religion, Christianity in particular.

Reading Nietzsche, Duncan believed she had found someone who understood what she was trying to do. From then on, she carried Nietzsche's books with her whenever she traveled, and quoted from them in speeches and essays. She referred to *The Birth of Tragedy* as "my Bible" (AD 107), and described *Thus Spoke Zarathustra* as "filled with phrases about man in his dancing being" (AD 123). She identified her own vision with that of Zarathustra: her dancing would help persons, especially women, overcome their faith in an otherworldly God by educating them into an awareness of their own bodily being as *holy* and *beautiful*, as the source of their highest ideals. For Duncan, dance is "religious art" in so far as it symbolizes the "freedom of woman" (IS 48).[9]

Martha Graham (1894–1991) presented her first concert of original choreography in 1926, the year before Duncan died.[10] She considered Duncan, alongside Ruth St. Denis, a mentor—one of the great matriarchs of modern dance. She first read Nietzsche nearly twenty years after Duncan had while on tour as a member of the company Denishawn, founded by St. Denis and her husband and dance partner, Ted Shawn. In one of her first interviews as a Denishawn dancer, Graham remarked: "I owe all that I am to Nietzsche and Schopenhauer" (Stodelle 1984:38).[11] The vision for dance that subsequently propelled her away from Denishawn was fueled by this admiration for Nietzsche. One of the first dances in which she developed her original aesthetics, titled "Dance" (1929), was accompanied in the program by a quotation from *The Genealogy of Morals*—"strong, free, joyful action." Like Duncan, Graham included Nietzsche's works on the reading lists she gave her dancers, and she kept copies of a number of his works in her personal library. Although she stopped referring directly to Nietzsche's works during the late 1930s, as the Nazis rose to power, she continued to speak of dance in ways that demonstrated his influence on her work. In a phrase she repeated throughout her career, Graham asserts that dance "is an affirmation of life through movement" (Armitage 1937:102–3).

In a manner similar to Duncan, what Graham found in Nietzsche was inspiration for *dancing religion*—that is, for presenting dance, in her dances, as capable of revaluing the hostility toward bodies, female bodies in particular, that she believed Americans had inherited from Christian belief and practice. Most of Graham's nearly 200 dances feature a religious figure (such as the Virgin Mary, Mary Magdalene, Eve, Lilith, Judith, Joan of Arc); enact a story drawn from a religious tradition (as in "Witch of Endor" (from the book of Samuel), "Song of Songs," "*Acts of Light*," "Embattled Garden" (Genesis)), or explore the terrain of religious themes and symbols (as in "Heretic," "Lamentation," "Diversion of Angels," "Dark Meadow"). The "religion" in question is almost always Christian; the central figure in her dances is almost always a woman. Moreover, the language Graham used to describe her dancing, whether speaking or writing, is saturated with religious vocabulary. She used religious language when teaching classes, and when coaching her company toward performance.

As with the dance imagery in Nietzsche's revaluation of Christian values, so too with the religion imagery and references to Nietzsche in Duncan and Graham's dance practice and performance: few scholars have asked the question why. While both women are recognized as giants in dance history, the *religious impulses* in their work—the language, themes, symbols, and stories they use in teaching, making, and performing their dances—have yet to receive the attention their ubiquity deserves.[12] Why did Duncan and Graham dance religion to the extent they did? What impact did reading Nietzsche have on the ways in which they engaged religious materials? What significance did these religious impulses have in the context of their respective processes of envisioning, practicing, making, and performing *dance*?

As is the case with Nietzsche's dance imagery, the reasons that scholars have sidelined these aspects of Duncan and Graham's work are woven into the history of modern dance and dance scholarship. Generations of dancers followed Graham and Duncan by doing what they did: rejecting and reinventing what had come before. Duncan and Graham (at least in the early days) rejected the codes, conventions, and techniques of classical ballet; they rejected the anti-dance values of Christian morality. They perceived themselves as rescuing dance from subservience to other arts. Subsequently, those who followed in their footsteps sought to "free" dance even further from "religion," from the burden of carrying symbolic or narrative meaning. In honoring Duncan and Graham's opposition to anti-dance values, later dancers rejected the religious impulses of Duncan and Graham's work as accidental and even detrimental to the practice and performance of "dance." This modernist impulse peaked in the 1960s in the postmodern experiments of Judson Church, when dancers performing pedestrian movements deconstructed the conventions of staging, timing, spacing, and training characteristic of concert dance.[13]

In the years following the rise of postmodern dance, dance history and scholarship blossomed. Dance studies shared with these movements the deconstructive eye that allowed one to take dance apart, analyze its constitutive elements, and thus develop conceptual frameworks for interpreting how dances make meaning (Foster 1986:chapter 4). In its development, dance scholarship also drew inspiration from cultural studies and cultural anthropology, areas of scholarship deeply influenced by Marxist theories of religion as ideology. Thus, dance scholars have tended not to treat the early modern dancers' religious themes and claims as adding anything to their processes of envisioning, practicing, making, and performing dances. And while scholars note the influence of Nietzsche on the advent of modern dance, they tend not to elaborate on the significance of dance imagery in his work or the role that imagery plays in his revaluation of Christian values. These tendencies have been further amplified by dance scholars' concern with establishing their field in the academy as having scientific status and a valid object of study, namely "dance." Dance scholars have responded to these challenges by conceiving dancing as a kind of writing, a language or text that can be read and interpreted to some extent.[14]

In these ways, then, the dearth of attention to the religious impulses of early American modern dance in dance scholarship may be linked to the dynamic that guides scholars in philosophy and religious studies to ignore dance imagery in Nietzsche's writing. In both fields, scholars remain bound by a conception and practice of *writing* that leads them to perceive dancing and religion as distinct and even antithetical activities in human life. This notion has prevented scholars from appreciating the significance of dance imagery in Nietzsche's revaluation of values, and from appreciating the art of Duncan and Graham as participating in cultural conversations concerning the nature and purpose of religion. As a result, scholarly work on Nietzsche, Duncan, and Graham has served to replicate the opposition between religion and dance that these three figures sought to dismantle as detrimental to the health and well-being of those in western culture. The dominant attitude toward religion among dance scholars and toward dance among Nietzsche scholars remains one of distrust and suspicion.

By reading Nietzsche, Duncan, and Graham together, then, *Nietzsche's Dancers* aims to address gaps in and between these fields of scholarship. *Nietzsche's Dancers* draws from the texts of Nietzsche to illuminate the theories and practices of the modern dancers; and draws on the writing and dancing of the modern dancers to illuminate Nietzsche's theories and practices. *Nietzsche's Dancers* argues that Duncan and Graham, by appropriating religious themes and describing their dance process in religious terms, were not only ruffling the white, Protestant, male clergy who published books against the immorality of social dancing,[15] they were also demonstrating what Nietzsche's project of revaluing values entails: challenging (Christian) belief in verbal expression as the privileged vehicle for communicating truth, knowledge, and ultimately, the Word of God. Duncan and Graham were developing dance as a practice for generating ideas about what "religion" is and should be. In reading these figures together, then, this book aims for more complete understandings of Nietzsche, Duncan, and Graham each in their own right, and in so doing, seeks to establish the significance of dance for the philosophy of religion.

Dance and the Philosophy of Religion

What I intend to offer through the course of this book is a case study in which dance proves relevant to the philosophy of religion. I intend to demonstrate, in imaginative recreations of their work, how Duncan and Graham engage and critically advance Nietzsche's project of revaluing all values, most notably Christian values toward embodiment. I do not claim that Duncan or Graham were intending to *do* philosophy; they were aiming to make good, significant art dance. However, in making art they deemed "good," I argue, they found it necessary to resist and reconfigure the Christian attitudes toward embodiment they found embedded in western culture. In so doing, their work offers valuable resources for philosophers engaged in theoretical discussions concerning the nature of "religion."

In presenting Duncan and Graham as case studies in Nietzsche's revaluation of values, my project is critical, creative, and constructive. As *critical*, it offers incisive analyses of how these figures engage the concepts of "dance" and "religion" in their writing and dancing so as to deflect mainstream Christian values toward bodies evident in the modern west. As *creative*, this project uses these analyses to imagine an ideal of dance in relation to religion which reading these figures together makes possible. As *constructive*, this project develops theoretical concepts alongside this vision which are capable of facilitating the further study and creation of phenomena recognizable as both religion and dance. As such this project is case specific: it does not offer a general theory of religion and/or dance. Nor is it theological: it does not offer a defense of dance as religious within the terms of a given tradition. Rather, it draws from the work of Nietzsche, Duncan, and Graham a vision for dance as a practice in which citizens of the modern western world come to know and value their bodies as processes of creative becoming, and thus sites for generating ideas of and about "religion."[16]

In this project, I draw inspiration from Grace Jantzen. Jantzen defines the task of the philosophy of religion as one of encouraging the full flourishing of human beings. It is a process she calls, borrowing from Luce Irigaray, "becoming divine" (1999:6). In Irigaray's words: "to become divine men and women, to become perfectly, [is] to refuse to allow parts of ourselves to shrivel and die that have the potential for growth and fulfillment" (Irigaray 1993b:68). In order to enable this flourishing, Jantzen contends, philosophers of religion must hold religious symbols as well as social and political systems accountable to the conditions such flourishing requires. They must follow in the footsteps of Zarathustra himself, rethinking the relationship between religion and secularism, and recreating a sense of "divine horizon" (1999:14). Women in particular must project images of the divine according to their gender, for dominant images of divinity in western culture have represented the experiences and interests of a system in which men occupy the highest echelons of power. With these ends in mind, Jantzen offers a *reconsideration* and *reconstruction* of the philosophy of religion as practiced in the Anglo-American tradition (3).[17]

Although Jantzen never mentions dance per se, her vision of philosophy of religion as a task of reconsidering and reconstructing images of divinity so as to enable human flourishing opens space for considering "dance" as a powerful ally in the philosophy of religion. Specifically, as their affinity with Nietzsche attests, Duncan and Graham were keen to enact their own reconsideration and reconstruction of western Christian values (10).[18] They did so not by using dance to project images of divinity per se, but by creating dance techniques and choreographies that would encourage people to appreciate their own bodily movement as the medium through which they exercise their creativity in generating and becoming the images of divinity they hold to be true. Duncan and Graham realized a vision for dance which I term *theopraxis*.

A Practice of Understanding

My method in presenting this case study involves an oscillating movement between Nietzsche's philosophy and the modern dancers' art that may be best described as *imaginative empathy*. Imaginative empathy refers to a practice of understanding. It is what Gerardus van der Leeuw develops at the core of his phenomenology of religion as an "indirect method" for discerning the meaning that phenomena have for those to whom they appear as "religious" or "religion" (LaMothe 2004:chapter 6). Van der Leeuw's method and my interpretation of it may both be read as responses to Nietzsche's call for *physiologists*—for readers who will seek out the physiological conditions that enable any cultural expression whether belief, idea, or value, to have the meaning it does for someone.

In relation to Nietzsche, Duncan, and Graham, I employ imaginative empathy in an effort to understand the meaning that their references to dance and religion have in the context of their work. I do so by recreating for myself the meaning that I perceive these references have for them. In other words, I do not claim to know what Nietzsche, Duncan, and Graham actually meant by their references. I do not claim to be able to experience what they experienced. All I know is what I can gather from my attempt to recreate for myself and others *my* experience of empathizing with the traces they have left behind in writing and dance.

To this end, I map the references to dance across Nietzsche's work and ask about the physiological conditions that enable these images to convey *meaning*. What experiences of dancing do these images presuppose and thus implicitly, if not directly, encourage among those seeking to overcome hostility to the body and affirm life? Conversely, I map the references to Nietzsche and religion across the work of the modern dancers—in their writing and speaking, their techniques of training, their dances themselves, and their theories of performance—and ask about the philosophical and theological problems that these references advance. What experiences of religion do these dances represent as possible and desirable? In dancing religion, what claims are these women making about what dance can and should be?

In these ways, my project is *empathetic* in the sense that in every moment, I am both discarding my own experience in an effort to step inside theirs, and drawing on everything I know in order to grasp what appears to me—including in the case of Nietzsche, my understanding of Duncan and Graham, and in the case of Duncan, my readings of Nietzsche and Graham, and in the case of Graham, my readings of Nietzsche and Duncan. The project is *imaginative* in that I aim to recreate what appears to me in written images that are able to communicate some sense of my experience of understanding to others. As both empathetic and imaginative, this approach is holistic, relational, and self-correcting: it engages a wide range of bodily experience, honors the nexus of relationships which those experiences represent, and respects the process of having and interpreting experiences as ongoing. Every insight I receive enables me to revisit a past insight anew in an ever-unfolding fabric of meanings. Using this approach, then, I imaginatively recreate the significance of Duncan and Graham's engagement with Nietzsche as representing their participation in the revaluation of Christian values.[19]

Chapter Outline

A brief outline of the argument that unfolds below in three parts and nine chapters will guide readers to the sections that promise the most immediate gratification, and hopefully thereby stir their interest in the rest.

Part I on Friedrich Nietzsche documents the range of dance imagery appearing across his corpus, discusses the influences that may have contributed to Nietzsche's interest in dancing, and offers interpretations of the most significant occurrences. Chapter 1, "First Steps," traces the dance imagery in two of Nietzsche's earliest writings, *Birth of Tragedy* (1967), first published in 1872, and *Human All Too Human* (1984), first published in 1878. In *Birth of Tragedy*, the dancing of the chorus is the key to the difference Nietzsche discerns between the value systems represented by Attic tragedy and Christian morality. The dancing of the chorus enables a tragedy to deliver an energizing encounter with the devastating paradox of Dionysian wisdom. A tragedy that does so catalyzes an affirmation or love of life in all of its manifestations. Even so, Nietzsche does not develop the implications of this insight about dance in this text. As he notes in his own "Attempt at Self-Criticism" written in 1886, he was prevented from doing so by his allegiances to Schopenhauer and Wagner such that in the critique of contemporary culture that concludes this book, he privileges music over dance.

With *Human All Too Human*, Nietzsche articulates his distance from the philosophies of Schopenhauer and Wagner, begins to make explicit his critique of Christian morality as hostile to embodiment, and introduces "dance" as an analogy for the kind of movement between science and art that "free spirits," including himself, practice. In doing so, this book sets up a parallel that Nietzsche later develops in the figure of Zarathustra: the free spirit as dancer. Based on an analysis of Nietzsche's theories of language and metaphor, this chapter argues that readers who interpret Nietzsche's dance images as metaphors for a changed mental state fail to take seriously his critique of belief as the defining locus of Christian religion.

Chapter 2, "Free Spirits," focuses on the pair of works that emerge together at the center of Nietzsche's corpus: *Gay Science* (1974), written in 1882 and 1886, and *Thus Spoke Zarathustra* (1954), written between 1882 and 1885. Book V of the *Gay Science*, written after Nietzsche finished *Zarathustra*, includes some of Nietzsche's most provocative statements about dance. Zarathustra himself is a dancer whose ability to dance signals his self-overcoming, his revaluation of Christian values, and his ability to affirm life in its most troubling expressions. Reading *Science* and *Zarathustra* through one another this chapter argues that Zarathustra, as a dancer, incarnates the faith in bodily becoming—the "great health"—that Nietzsche advocates in *Science*. Zarathustra not only teaches that the body is a process of its own creative becoming, a "great reason," a "doing 'I,'" he demonstrates it in his own self-overcoming. Dancing is, as in the opening quotation, Zarathustra's ideal, art, and piety. This chapter introduces the concept of *theopraxis* as a description of the vision for dance that Zarathustra realizes.

Chapter 3, "Loving Life" examines the dance imagery that appears in Nietzsche's works from 1886–1888. In these works, Nietzsche fleshes out his vision for "dance" as a figure for a physical–spiritual practice that serves to counter the effects of what he calls the "ascetic ideal." Nietzsche is increasingly clear in these writings about how the process of overcoming and revaluing values will require that people relearn their relationship to their bodily being. People must learn to embrace what he calls the "physiological contradiction" characteristic of those living in a modern time: we embody Christian values and a scientific resistance to them. The practice of dance, for him, represents the kind of movement in relation to oneself that allows a person to affirm this physiological contradiction as a condition enabling great health.

Part II introduces Isadora Duncan as someone who acknowledges her own participation in this physiological contradiction and engages dance as a medium for overcoming and revaluing it. Chapter 4, "A Dionysian Artist," explores Nietzsche's influence on Duncan's vision for what dance can and should be—a "high religious art," a catalyst for a rebirth of religion, particularly in its attitudes toward women (AD 62). It examines Duncan's account of herself as a "Pagan Puritan, or a Puritanical Pagan" (ML 255) before investigating the three principles she distills for guiding her in discovering a dance capable of overcoming this contradiction: studying nature, harmonizing movement and form, and awakening soul. This chapter then demonstrates how Duncan intended to realize these principles by *being the chorus* that Nietzsche describes in *Birth of Tragedy* (AD 96).

Chapter 5, "Incarnating Faith" builds on chapter 4 in investigating Nietzsche's influence on other moments of Duncan's dance process—her approach to dance education, her practice, her dance making, and her performance. This chapter argues that Duncan's goal was to develop a practice of dance that would help people learn to participate in the rhythms of their own bodily becoming. This chapter concludes with an assessment of how Duncan critically advances Nietzsche's work along a trajectory of her engagement with him in developing her dancing as a means for incarnating faith in a *female* body. In so doing she reveals why his project of revaluing Christian attitudes toward embodiment requires revaluing gender as only (her vision for and practice of) dancing can.

In Part III, Chapter 6, "An Affirmation of Life," argues that Graham's vision for dance as an affirmation of life through movement represents what she has learned from Duncan's experience about how dance can and should participate in the revaluation of Christian values Nietzsche describes. Chapter 6 investigates the traces of Nietzsche's influence on Graham's vision for dance as an affirmation of life through movement. For Graham, a dance that catalyzes such an affirmation is "significant movement" (Armitage 1937:98), or "movement made divinely significant" (G32:5), for the way in which it reveals *human being*. Along the way this chapter offers close analyses of *Dance, Heretic* (1929), and *Primitive Mysteries* (1931).

Chapter 7, "Athletes of God," investigates the Nietzschean influences on Graham's discovery and distillation of the contraction and release—her stylized movements of breathing—as the generative kernel of her dance technique. By practicing kinetic images of the movements of one's own breathing, this chapter suggests, a person comes to know her or his body as a dynamic, reflexive process of becoming. This chapter elucidates three kinetic images that the practice of Graham's technique draws into physical consciousness: the movements of contraction and release, center, and ground. For each, chapter 7 analyzes the religion language Graham uses to describe the kinetic image at hand, and in so doing explains how Graham intends the practice of her technique to participate in a Nietzschean revaluation of Christian values.

Chapter 8, "Words to Dance," investigates how Graham, based on what she learned from Duncan and Nietzsche, uses words and texts from religious sources in making her dances. Beginning with an analysis of Graham's attitudes toward words and books, this chapter examines the record Graham left behind in her *Notebooks* of how she read texts of and about religion. The chapter then explores the relationship between text and dance enacted in Graham's evening length masterpiece, *Clytemnestra* (1958), a dance inspired by one of the tragedies whose birth Nietzsche recounts, Aeschylus's *Oresteia*. In the dance Graham seeks to recreate the visceral impact of the text and in so doing reveal how this woman's tragic tale can catalyze an affirmation of *life*, that is, of a spectator's participation in the ongoing rhythms of bodily self-creation. The final sections of this chapter analyze two other dances by Graham, *Errand into the Maze* (1947) and *Acts of Light* (1981), in order to clarify and nuance the picture of how Graham's use of texts reflects critically and constructively on Nietzsche's revaluation of values.

The Conclusion, "Another Ideal," articulates an ideal of dance—a philosophy of dance in relation to religion—that I discern through my study of Nietzsche's dance images, and Duncan and Graham's appropriation of them. In this philosophy, dance appears as an ideal: for a practice that enables people to relearn their relationship to bodily being; for a knowledge of self as a bodily becoming; and for a moral perspective on religion that holds religious symbols and beliefs accountable to the love of self such practice and knowledge express.

Part 1

Friedrich Nietzsche

Often one could have seen me dance; in those days I could walk in the mountains for seven or eight hours without a trace of weariness. I slept well, I laughed much—my vigor and patience were perfect.

—K6, 341; EH 303

A body sits alone in a chair at a desk, writing. Feet are on the floor or crossed or curled under the thighs. Lower abdomen is slack. Tension creeps along the arms and into the shoulders and neck. Breathing is shallow. The head bends over the page. Eyes squint to concentrate the thinking that bubbles up within the body. As words cluster in the space behind the forehead, they are squeezed out through the tips of fingers and pen onto the page. The words represent this body that sits and thinks.

Or do they? The body that writes is also a body, at least in Nietzsche's case, that was ill. Throughout his life Nietzsche was plagued by symptoms whose cause has yet to be definitively determined. Beginning as a boy in school, he struggled with headaches and nausea. As a young adult his health worsened to the point that he resigned from his post as a professor at the University of Basel, at the age of 34. For the next ten years, he wrote in, through, and between bouts of sickness. Severe headaches, attacks of vomiting, and insomnia-induced exhaustion kept him in bed for days at a time. In response, he sought solitude, enabling climates, the right balance of foods, and restorative recreation.

Most of all whenever and wherever he could, Nietzsche walked. He walked on mountain paths, along lakes, through the woods; he walked along the streets of towns, in the chilly rain or warm sun. He walked for hours, not when he felt good, not because he felt good, but in order to feel good.

He walked to stir the sparks of his meager health into a flame. He walked to breathe, to think, and eventually, to write. When he hiked in the mountains he would have worked hard. Such walking is strenuous. The heart beats quickly, strongly; blood pounds through temples and limbs; muscles ache and release. The spine straightens after hours of curving over a manuscript. Eyes relax after straining to focus on tiny words. Legs unfold after curling and cramping. Breathing deepens and expands, filling the chest with lightness, air. A body opens to life. No longer a mind working feverishly against time and the limitations of the flesh, the self is also body. Walking, Nietzsche cultivated the kind of health—the *physical consciousness*—that enabled him to resist the values of Christian morality and embrace his own metaphor making power as a writer.

On days when Nietzsche's health overcame his sickness, he not only walked. He may even have danced. He danced to unfold and experience his inspiration: "the suppleness of my muscles has always been greatest when my creative energies were flowing most abundantly. The *body* is inspired" (K6, 341; EH 303). And when he felt the dance welling in him, Nietzsche loved life, all of it, even his sickness.

Perhaps the words on his page represent a body that not only writes but also dances.

Chapter 1

First Steps

By virtue of this idealizing art, some peoples have turned diseases into great beneficial forces of culture—the Greeks, for example, who in earlier centuries suffered from wide-spread nervous epidemics (similar to epilepsy and the St. Vitus Dance) and created the glorious prototype of the bacchante from them.

—K2, 174; HH #214, 128[1]

It is not immediately evident why Nietzsche would choose dance to play such a prominent role in his project of revaluing all values. Nietzsche never studied dance as an *art*, which at the time meant classical ballet. It is not even clear that such study would have been possible as Germany had not yet developed its own ballet culture. He may have danced folk dances when vacationing with his cousins at his mother's family estate; he may have marched and paraded with the boys at his boarding school, the prestigious Pforta Academy, during their yearly festivals; he may also have learned some social dances such as the newly popular waltz as a student in university.[2] He had at least heard about such dances from his sister Elisabeth, younger by two years, who regaled him with tales of dancing lessons at her finishing school.[3] He also would have learned the forms of folk and social dances through the love of classical music he shared with his father who died when he was four. Yet the young Nietzsche remained content to play the music on the piano, improvising for friends and family and composing his own pieces.

Even if taking classes in dance had been possible, it is not clear that Nietzsche's mother would have allowed it. Dancing was not a proper occupation for a Christian child, and certainly not for a boy destined to become a pastor in the Lutheran Church. Nietzsche's father had been a Lutheran pastor, as had his paternal and maternal grandfathers. Born in the parsonage of the small town of Röcken, Nietzsche moved after his father died, along with his mother, his grandmother, his father's two unmarried sisters, and Elisabeth, to the medieval, walled city of Naumberg. These five pious women hoped that he would be the next family cleric. While the Lutheran tradition did welcome music as an integral part of liturgical practice more than a number of other traditions that emerged in the wake of the Protestant Reformation, Nietzsche would never have been expected to dance or to watch dance in church.

Nietzsche's primary exposure to dancing, then, seems to have come through reading—the Bible, romantic literature, and the classics of Greek and Roman antiquity. From a young age Nietzsche steeped his imagination in biblical verses, and continued to do so as a student of Christian theology. The Bible is replete with passages in which dance is embraced, encouraged, and even required as an expression of relationship with God, and there is evidence that Nietzsche noticed. A number of Nietzsche's references to dance, for example, allude to the passages from the New Testament in Matthew and Luke where Jesus remarks: "But to what shall I compare this generation? It is like children sitting in the market places and calling to their playmates, 'We piped to you and you did not dance; we wailed, and you did not mourn' " (Mt 11:16–17).[4] Though Nietzsche often conflates a piping Jesus with the pied piper and the flute-playing Greek god Dionysus, the biblical critique resonates through his usage: a person who does not dance is one who does not have *faith*, one who refuses to hear the music of life, or participate in the rhythms of creation. In addition, a number of Zarathustra's speeches echo the Psalms, including the speech in the Fourth Part, "On the Higher Man," in which Zarathustra, like the psalmist, urges his listeners to lift up their legs and dance.[5] Thus even though his Lutheran tradition would not countenance dancing in church, Nietzsche's Christian heritage offered authoritative cases in which dancing served as a medium of religious experience and expression.

Additional influences on Nietzsche's use of dance imagery may have come from his studies of romantic art, philosophy, and poetry. Nietzsche warmed to the writings of poets and philosophers in the early nineteenth century who embraced art as a medium of revelation over and against the clear reason of science.[6] Although dance, when it appeared in discussions of art, regularly ranked below the fine arts of poetry, music, and visual art, it did appear, and often in a context evoking alternative traditions to those of modern Christianity. In the 1830s, moreover, as ballet entered its classical period, romantic literature proved a primary source for narratives, symbols, characters, and themes. The first ballet to appear in the newly constructed Paris Opera in 1830, "Le Dieu et la Bayadere," was based on a poem of the same name by Goethe. A heady mix of sensuality, oriental exoticism, and bourgeois morality characterized this ballet and others like it; the dance technique emphasized speed, agility, and extension.[7] The star of the performance was always a woman whose impersonation of ethereal creatures—nymphs and sylphs as well as prostitutes or temple dancers—belied the tremendous strength and discipline required to sustain her illusion of buoyant effortlessness. Images of the most famous of these ballet dancers feature feet that dwindle to a point. Such images trained spectators' attention to what the dancers sought to perfect: a sense of weightlessness. It is hard to imagine that Nietzsche's repeated references to light feet and the spirit of gravity were not influenced by these aesthetics of classical ballet.

At the same time, while Nietzsche celebrated the romantic movement's embrace of art, illusion, and sensuality, he decried what he called its *decadence*—a decadence nowhere more pronounced, perhaps, than in the narratives and aesthetics of classical ballet. Decadence, for Nietzsche, involves, among other things, using sensuality to support Christian morality rather than oppose it. Nearly every romantic ballet narrative did exactly this, indulging in an otherworldly fantasy before returning to affirm Christian, bourgeois, moral values. Although nowhere in his published works does Nietzsche engage in a sustained critique of classical ballet, he does discuss Richard Wagner at length as a primary example of romantic decadence. Wagner was a man and musician whom Nietzsche loved and revered as a father, and an artist whose vision for opera always included a ballet.

In the end, it was in Nietzsche's studies of Greek and Roman antiquity—their philosophy, epics, poetry, and drama—that alongside the Bible seem to have exerted the greatest influence on his conception and use of dance imagery. Nietzsche's classical studies took root and flourished during his six years at Pforta. In this all-male establishment, run with a blend of monastic and military rigor, Nietzsche disciplined himself to a daily regime that included mandatory chapel, communal meals, hours of classes, and mere minutes of unstructured time. In the course of a day he moved through dimly-lit buildings whose austere medieval architecture reflected the cool and cloudy climate of northern Germany. Practices designed to exercise the mind and still the body found justification in the Christian morality preached there.

In these halls, often squinting to see the words on the page, Nietzsche delved into the sun-strewn world of southern Europe depicted in ancient texts—a world whose climate he would later seek out as most beneficial to his health and his writing. The contrast between the ancient world and his Christian surroundings could not have been more stark. The differences between light and dark, warm and cold, open and cloistered impressed Nietzsche as signaling differences in morality and aesthetics, theology and metaphysics. In the Greek and Roman texts gods and goddesses not only transgressed principles of Christian morality, they danced, as did the devotees who worshipped them. The juxtaposition of these worlds may have provided Nietzsche with the perspective from which to begin asking questions about his commitments to Christianity, to the Lutheran tradition, and to his family's vision for his future.

In the year following his graduation from Pforta, the tension in him between the competing modes of valuation represented by these two worlds erupted. Finding it increasingly difficult to support the exclusivity of Christian claims, Nietzsche shifted his course of study from theology to philology. As commentators note, Nietzsche may have developed a rational critique of Christian faith upon reading Ludwig Feuerbach in 1861.[8] Feuerbach had developed an early theory of religion as a function of *projection*: humans project perfected forms of their own qualities onto an ideal they call "God."[9] Feuerbach sought to discern the truth of classical expressions of theology by reversing them, interpreting "God created human" to mean that "human created God." However, Nietzsche did not leave theology to study philosophy with its principles of reason and logic. His antidote to Christian theology was classical philology, and then, not solely the study of Socrates' clear reason, but the study of poetry and drama as well. It was in philology that Nietzsche found resources for entering a world in which *art* is inseparably entwined with *religion*—a world in which *dance* plays an integral role in the moral life of the community. Nietzsche's shift from theology to philology suggests that his opposition to Christian morality did not take root in the field of rational argument. Rather than refute Christian values, Nietzsche intended to *revalue* them—that is, to examine them, reveal their contradictions, and in doing so, give rise to alternative ideals. Only through such revaluation, as Nietzsche develops, can we hope to *overcome* the influence of Christian values on our thinking, feeling, and acting.

In 1871, recuperating from injuries sustained in voluntary military service during the Franco-Prussian War, Nietzsche wrote the book in which he first articulates for himself what the culture of ancient Greece meant to him as "the only parable and parallel in history for my own inmost experience" (K6, 311; EH 271). In particular, the Attic tragedy of Greece, fifth century BCE, provided such a parable and parallel in offering him "a formula for the highest affirmation [*der höchsten Behajung*], born of fullness, of overfullness, a Yes-saying [*Ja-sagan*] without reservation, even to suffering, even to guilt, even to everything that is questionable and strange in existence. Nothing in existence may be subtracted, nothing is dispensable" (K6, 311; EH 272). This formula for the highest affirmation represented to him what he found lacking in Christian values, and thus a perspective from which to critique them.[10] Where Christian authorities oppose morality, as the bastion of truth, to art, sensuality, and bodily life as the realm of lies, Nietzsche observes a "degenerating instinct that turns against life" (BT 23). While he admits the "careful and hostile silence with which Christianity is treated throughout the whole book" (K1, 18; BT 23; also K6, 310; EH 271), it is also clear that his exposition of Attic tragedy in *Birth* correlates point by point with those aspects of Christian teachings he perceives as failing to cat- alyze an affirmation of life.[11] In this exposition, it is *dancing*—the dancing of the chorus—that appears as the element that distinguishes a Christian "doctrine and valuation of life" from one that is "purely artistic and *anti- Christian*": "What to call it? . . . Dionysian" (K1, 19; BT 24).

Even though Nietzsche describes *Birth* as his first revaluation of values, it was not until years later that he discerned the significance of what he had accomplished—and its limitations. In his "Attempt at Self-Criticism," a preface composed fourteen years after the initial printing, Nietzsche is particularly critical of himself first, for lacking the "courage (or immodesty?) to permit myself in every way a language of my own," (K1, 24; BT 21) and second for having distorted or "spoiled" the Greek problem by harnessing it to an interrogation of German culture, and the music of Wagner in partic- ular. Still under the influence of Schopenhauer, Wagner, and Kant, Nietzsche had concluded his essay by privileging music above the other arts as harboring the potential for a rebirth of Dionysian art in contemporary culture. The problem: dance effectively drops out of his analysis. In the preface, Nietzsche's effort to remedy the situation appears in his references to Zarathustra. Instead of heralding a musical Socrates as his vision for a counter ideal to Christian morality (as he does in the final sections of *Birth*), he here wreaths a dancing Zarathustra. In quoting "that Dionysian monster who bears the name of Zarathustra," Nietzsche selects a passage from "On the Higher Man" in which Zarathustra urges the higher men to dance and describes himself as "Zarathustra, the dancer; Zarathustra the light one who beckons with his wings, preparing for flight" (K1, 22; BT 27).

While commentators often follow Nietzsche's lead and attribute less importance to section 16 and those that follow, the relationship between the first and second parts of *Birth* is of utmost importance in understanding the rise of dance in Nietzsche's project as an image for the revaluation of Christian values. A crucial juncture in that rise occurs in Nietzsche's *Human All Too Human*, where he not only articulates his difference from Schopenhauer, Wagner, and Kant and finds his "own lan- guage"; but also introduces "dance" as "an analogy" for what free spirits do when they affirm and love life in all of its manifestations. In weaving among the various references to dance in *Birth* and *Human*, this chapter produces a web whose sticky threads capture a seldom-noticed dimension of Nietzsche's work: namely, his belief that the process of revaluing values requires that a person engage in practices that enable him or her to develop a new relationship to his or her own bodily being, one akin to that catalyzed in the spectator by the experience of watching the chorus of Attic tragedy dance.

A Magic Transformation: *The Birth of Tragedy*

[M]agic transformation [Verzauberung] *is the presupposition of all dramatic art. In this magic transformation the Dionysian reveler sees himself as a satyr, and as a satyr, in turn, he sees the god, which means that in his metamorphosis he beholds another vision outside himself, as the Apollinian complement of his own state. With this new vision the drama is complete. (K1, 61–2; BT 64)*

Nietzsche's approach to tragedy in *Birth* was not typical of a philologist or a philosopher. Rather than treat a tragedy as a text he approached it as a performance art; rather than seek in that text aesthetic values, he sought answers to issues of metaphysics and morals that contrasted with the Christian responses with which he was familiar. Rather than ask about the meaning of the story, he posed the question *how*: how is it that the performance of an Attic tragedy effects an affirmation of life in spectators? That it did effect such affirmation, he assumes.

That the *dancing* of the chorus in Attic tragedy holds the key to understanding its efficacy appears in Nietzsche's descriptions of the "magic transformation" that tragedy catalyzed in spectators. Such a transformation can occur, according to Nietzsche, when a work of art succeeds in negotiating a reconciliation, however temporary, between two artistic energies—energies of individuation and dissolution, creation and destruction—to which he gives the names "Apollinian" and "Dionysian," respectively. Nietzsche roots each of these energy-metaphors in a physiological state. He associates the Apollinian with a dream-like clarity and calm and the Dionysian with a dizzying intoxication or rapture [*Rausch*]. While both energies are necessarily present in every work of art, Nietzsche admits that one energy may dominate, as Apollinian energy does in the visual arts, or Dionysian energy in the arts of music and dance.

Despite his professed silence regarding Christianity, the first example Nietzsche gives of a Dionysian phenomenon is, ironically enough, Christian. He writes: "In the German Middle Ages, too, singing and dancing crowds, ever increasing in number, whirled themselves from place to place under this same Dionysian impulse. In these dancers of St. John and St. Vitus, we rediscover the Bacchic choruses of the Greeks, with their prehistory in Asia

Minor, as far back as Babylon and the orgiastic Sacaea" (K1, 29; BT 36).
In introducing this case, Nietzsche traces the hinge of an implicit critique
that runs through the book: Christians share with the Greek world participation in Apollinian and Dionysian impulses they simultaneously deny.[11]
When Christians interpret Greek culture as pessimistic or barbaric, or when
they interpret the Dionysian impulses in their own tradition as evidence of
sickness, what they are expressing is a misunderstanding of their own physiology. Their interpretations bear witness to their own impoverished life
condition. The difference between Christian and Dionysian is not, strictly
speaking, an opposition. It represents a degree of health.

Setting the stage for the impending contrast, Nietzsche proceeds to pin
the genius of Attic tragedy on the degree of cooperation between these two
physiological energies that is established by the relationship between the
dancing and singing chorus, on the one hand, and the dramatic narrative,
on the other. Commonly, scholars identify the chorus as Dionysian and the
drama as Apollinian; yet the genius of Attic tragedy as Nietzsche describes
it is the degree to which these energies interpenetrate in all moments of the
tragedy. As he explains, the Apollinian narrative tells a *tragedy*. Even though
the drama provides comforting dream-like images of individuals who speak
and act, its *message* is Dionysian: "the shattering of the individual and his
fusion with primal being. Thus the drama is the Dionysian embodiment of
Dionysian insights and effects" (K1, 62; BT 65). Conversely, even though
the chorus offers an intoxicating mix of elemental rhythms, kinetic and
aural, that impel spectators to identify viscerally with the anonymous members of the chorus, those members also project a visual image. Not only do
the members dance and sing in unison, they appear in the form of *satyrs*:
"the force of this vision [*die Kraft dieser Vision*] is strong enough to make the
eye insensitive and blind to the impression of 'reality,' to the men of culture
who occupy the rows of seats all around" (K1, 60; BT 63). Given his visceral
identification, what the spectator sees in the satyr is *himself* as "the image of
nature and its strongest urges, even their symbol, and at the same time the
proclaimer of her wisdom and art—musician, poet, dancer [*Tanzer*], and
seer of spirits in one person" (K1, 63; BT 65–6).

It is this crossing of the energies, then, that renders Attic tragedy so effective: it cements a set of visceral and visual identifications between spectator
and performance that provide a spectator a perspective from which to
perceive the tragic narrative as an inducement to bliss. In so far as the
elemental rhythms of the singing and dancing spark a *visceral identification*
of spectator with chorus, the spectator is drawn to see herself in the image
of a satyr and thus see herself in relation to the dramatic narrative on stage
as the agent through which the characters of that narrative come to life. She
is able to sense in that narrative a power of nature of which she is a member,

by which she is created, and in whose power she herself participates as an image-creator. In this one moment, in other words, she experiences the two faces of Dionysian wisdom: as an individual actor she is doomed; as a member of the chorus she is ecstatic, immortal. In short it is the Dionysian message of the *narrative*—"the terrible destructiveness of so-called world history as well as the cruelty of nature" (K1, 56; BT 59)—*combined* with the Apollinian image of the dancing *chorus* that precipitates the transformation.

As the spectator emerges from this experience, he does so with an ability he did not have before: an ability to affirm or love life. Nietzsche summarizes:

> We are really for a brief moment primordial being itself [*das Urwesen selbst*], feeling its raging desire for existence [*Daseinlust*] and joy in existence; the struggle, the pain, the destruction of phenomena, now appear necessary to us, in view of the excess of countless forms of existence which force and push one another into life, in view of the exuberant fertility of the universal will . . . In spite of fear and pity, we are the happy living beings, not as individuals, but as the *one* living being [*Lebendige*], with whose creative joy [*Zeugenslust*] we are united [*verschmolzen*—fused, merged]. (K1, 109; BT 104)

It is because the spectators have suffered the sting of the tragic narrative that they are able to experience themselves as "happy living being." The elemental rhythms of singing and dancing release in them a strength to affirm that sting as evidence of their participation in the "eternal joy of becoming": the "power" of the "moods and insights" they experience "transforms them before their own eyes till they imagine that they are beholding themselves as restored geniuses of nature, as satyrs . . . everything is merely a great sublime chorus of dancing and singing satyrs or of those who permit themselves to be represented by such satyrs" (K1, 59; BT 62). In so far as a spectator permits himself to be represented by such satyrs, he comes to know that he is necessary to life not only as one of its countless forms, but as an image-maker himself.

In sum, Attic tragedy, as a performance art, stages a moment in which a spectator experiences a horrifying account of suffering as a stimulus to loving life. It does so by occasioning a *transformation* in a spectator's *experience* of herself, her world, and her relationship to what appears to be. This transformation is primarily *physiological*. The parts of the tragedy establish patterns of *visceral* and *visual identification* with spectators such that the spectators feel and see themselves, in their own bodily being, as participating creatively in an on-going, all-powerful rhythm of creation and destruction. A spectator feels welling within her a capacity to *affirm life*—not for any *reason*, nor through an act of will, but as a necessary expression of an overwhelming sense of well-being, gratitude, and love. Said otherwise, she does not think or reason that joy is deeper than suffering; she *feels* it. She feels it in her own body as a sense of energy and health, as her own truth. The performance thereby reeducates a spectator's senses, opens her to new possibilities for perception, and provides her with grounds for generating values that affirm life.[12]

Dance Revealed

What few notice in Nietzsche's account of this magic transformation is the indispensable role played by the dancing of the chorus in mediating the paradoxical effects of suffering and bliss. It is the *dancing* that conducts the kind of feeling most essential to this transformation based on the way in which dancing alone among the arts animates the *body*. Dancing, for Nietzsche, is first and foremost a *symbolic* activity, a form of human expression. As a form of expression, it is unique in that it makes use of the body as a whole—and not just parts of the body employed in language or music or even drama.

That dancing engages the body, or bodily movement, as the medium of expression is what enables the dancing both to establish a visceral connection between chorus and audience and communicate across that identification a sense of Dionysian rapture. Describing the eruption of Dionysian energies in Greek culture, Nietzsche writes: "we need a new world of symbols; and the entire symbolism of the body [*die ganzeleibliche Symbolik*] is called into play, not the mere symbolism of the lips, face, and speech but the whole pantomime of dancing, forcing every member into rhythmic movement. Then the other symbolic powers suddenly press forward, particularly those of music, in rhythmics, dynamics, and harmony" (K1, 33–4; BT 41). Full-body symbol-making is here celebrated for its ability to express the "essence of nature" and it does so by enabling people to feel their bodily selves as subsumed and animated by a power greater than themselves. In this regard, dancing, or body symbol-making represents the fullest unfolding of a person's creative capacities; other arts, including music, emerge in the wake of dancing as representing further constraints on bodily movement.

The effect of this creative experience has profound implications for Nietzsche. The experience of dancing allows a person to sense and experience his own body differently. In the act of dancing, he is not only a body; he is a body making images of itself, kinetic images. He is a body becoming the kinetic images he makes of himself as a member of nature. And in so far as his moving body produces kinetic images that stir others to move, he knows his own body—in its capacity to create kinetic images of itself—as participating in the rhythms of creation and destruction that life or nature is. "Singing and dancing [my translation]" Nietzsche writes,

> man expresses himself as a member of a higher community; he has forgotten how to walk and speak and is on the way toward flying in the air, dancing [*tanzend in die Lüfte emporzufliegen*]. His very gestures express enchantment. Just as the animals now talk, and the earth yields milk and honey, supernatural sounds emanate from him, too: he feels himself a god [*als Gott fühlt er sich*], he himself now walks about enchanted, in ecstasy, like the gods he saw walking in his dreams. He is no longer an artist, he has become a work of art: in these paroxysms of intoxication [*Rausch*] the artistic power of all nature reveals itself to the highest gratification of the primordial unity. The noblest clay, the most costly marble, man, is here kneaded and cut. (K1, 30; BT 37)

Here the experience of dancing transforms: *he feels himself a god*. His own gestures and the sounds they evoke from him give him a sense of himself as empowered. He senses his bodily self as an artist making kinetic images and as the god those images represent him to be, creator and created. With each gesture, moreover, he confirms this knowledge: he becomes the person capable of making and being the godlike creature his gestures represent. In short, dancing, as a bodily symbol-making, is always (more than) double. A dancer is subject and object, thing-in-itself and appearance, and the movement that gives rise to both in their relation to one another.

Thus, in Nietzsche's account of Attic tragedy, the act of dancing provides an ingredient that no other art, not even music, can provide: an experience of bodily movement as *becoming*, as a two-fold movement, and thus, as the medium in and through which *values* become real or incarnate. Spectators of a tragedy, even if they do not actually dance, register the kinetic effects of the movement through their visceral and visual identification with the choral members. Responding involuntarily to the rhythmic forms, spectators usher in a sense of themselves as images and image-makers. In this way Attic tragedy communicates what I call a *physical consciousness*: a bodily sense of my body as a process of its own becoming. It does so by exercising my capacity for making *kinetic images*—a body's images of itself that appear only in the making of the movement patterns they enable. From the perspective of this physical consciousness, the nausea of knowing that I owe my individuality to forces of creation and destruction that are indifferent to my existence is tempered by the knowledge that even this image of nature is a function of my participation in it. In turn, my sense of individual mastery and agency is tempered by an experience of that individuality as an illusion in whose creation I participate.

Nietzsche relays this paradox in one of the more opaque sentences in *Birth*: "[T]he dialogue is an image of the Hellene whose nature is revealed in the dance because in the dance the greatest strength remains only potential but betrays itself in the suppleness and wealth of movement" (K1, 64; BT 67). The dancing of the chorus "reveals" the nature of what the "image of the Hellene" in the dialogue also represents. It does so by "betraying itself," that is, by offering rhythmic forms and kinetic images whose clarity, pattern, and intricacy suggest a potential, a "greatest strength"—a suppleness and wealth that overflows in an ability to experience the suffering in the drama as a stimulus to a greater affirmation and love of life. As a stimulus to dance.

A Musical Socrates

Given the significance of dancing in Nietzsche's account of Attic tragedy, it is surprising that dance seems to disappear from his discussions when he moves in the later sections of *Birth* to claim the title of Dionysian art for German music, Wagner's in particular. Almost everywhere Nietzsche mentions the chorus, he refers only to the singing, the music. In doing so he also makes claims that nearly contradict the points discussed earlier regarding the interdependence of music and dance.

The influence of Schopenhauer on this shift in focus is evident. As Nietzsche himself admitted, he bends his own analysis of Attic tragedy to support Schopenhauer's claim that music, over and against all of the other arts, bears a unique relation to what he calls the "will," Kant's thing-in-itself. For Schopenhauer, the will, as "a craving, desiring, wishing, or a detesting, shunning and not wishing, is a feature of every consciousness: man has it in common with the polyp" (Schopenhauer 89). The will gives rise to all manifestations of life. Although Schopenhauer's argument initially promises to embrace "the body" as the means by which a person knows the will, he eventually concludes that the will is insatiable, and that, as a result, a human's only hope for peace and well-being is to rise above its endless striving in aesthetic contemplation (120). In such a state we are completely absorbed into the moment, we forget our subjectivity, our embodied particularity; we are cheered and calmed (120–1). It is music, he insists, in contrast to the other arts that provides humans with this opportunity. Hearing has an immediate connection with the will; and music provides direct access to feelings and thought without the interference of the body (168). Music, Schopenhauer concludes, is a copy of the will itself (164).

Nietzsche's attempt to align his analysis with Schopenauer's appears in passages like the following: "it is only through the spirit of music that we can understand the joy involved in the annihilation [*Vernichtung*] of the individual. For it is only in the particular examples of such annihilation that we see clearly the eternal phenomenon of Dionysian art, which gives expression to the will in its omnipotence, as it were, behind the *principium individuationis*, the eternal life beyond all phenomena [*Erschienung*], and despite all annihilation" (K1, 108; BT 104). By this account music *alone* accomplishes the effects that required the cooperation of dance in the earlier sections. It is music that enables us to understand and enjoy the "eternal life beyond all phenomena" because music communicates the annihilation of individual subjectivity in a Dionysian rapture and the release from the endlessly striving will.

Yet what is missing from this account is the dimension of Nietzsche's earlier analysis that described how a person comes to know her bodily being as the medium in and through which she participates in the creative rhythms of life. In upholding the exclusive rank Schopenhauer accords to music, Nietzsche unwittingly smuggles in a value that he writes the first section of the book to reject: an ascetic denial of the body. Nietzsche drives a wedge between the dancing and the singing of the chorus and focuses on the latter to the exclusion of the former.

As a result, in applying his analysis of Attic tragedy to the case of contemporary German culture, Nietzsche, as he later admits, distorts both the problem and the solution. In his initial analysis of Attic tragedy and its fall, he argued that tragedy loses its ability to catalyze an affirmation of life when Euripedes, under the influence of Socrates, creates a drama whose meaning lies in the moral of its dramatic narrative. Art is textualized, rationalized. As Nietzsche explains: Socrates peddled "a profound *illusion*" whose spell extends to the present day: "the unshakable faith that thought, using the thread of causality, can penetrate the deepest abysses of being, and that thought is capable not only of knowing being but even of *correcting* it" (K1, 99; BT 95). For those under the influence of this *Wahnvorstellung* or "hallucination, delusion," it is impossible to perceive the dancing or singing of the chorus as effective to the meaning of the experience. As Nietzsche contests, "we are reduced to a frame of mind which makes impossible any reception of the mythical; for the myth [*Mythus*] wants to be experienced vividly as a unique example of a universality and truth that gaze into the infinite" (K1, 112; BT 107). To experience the narrative vividly, as a myth, is to have one's experience of embodiment transformed by it such that it catalyzes awareness of one's bodily becoming in the ways described earlier.

Yet, in the end, Nietzsche makes the mistake he attributes to Socrates. Under the influence of Schopenhauer, after observing that this reduction of tragedy to text characterizes the contemporary German situation as well, Nietzsche calls for a rebirth of music only and not of dance. It is music that will restore the experiential dimension to the text's moral.[13] He writes: "music imparts to the tragic myth an intense and convincing metaphysical significance that word and image without this singular help could never have attained. And above all, it is through music that the tragic spectator is overcome by an assured premonition of a highest pleasure [*ein hochsten Lust*] attained through destruction [*Untergang*] and negation, so he feels as if the innermost abyss [*Abgrund*] of things spoke to him perceptibly" (K1, 134–5; BT 126). On the one hand, this account mentions experiential transformation as an essential element in understanding the impact of tragedy: a person feels and "is overcome." On the other hand, in so far as Nietzsche attributes that effect to *music* the scope he allots for the "magic transformation" shrinks. Here the spectator is overcome by a *premonition* of pleasure. She is what Nietzsche calls an *aesthetic listener*. Yet listening does not guarantee that a person will experience her bodily becoming as a locus for developing physical consciousness, exercising kinetic imagination, or incarnating values. The spectator is passive. The passivity of listening can even encourage the further entrenchment of the mistaken perception that we are minds in bodies, that the meaning of tragedy is in its text, and that the moral of the story is to deny and rise above the rhythms of will. In so far as a spectator does not experience the visceral identification of her bodily self with the chorus that would allow her to affirm her participation in the process of creation it represents, she experiences reality as an "innermost abyss" and not an endless fund of creative power.

Thus, when Nietzsche concludes *Birth* calling for a Socrates who will play music, he betrays his own insights. Where in Attic tragedy, the dancing mediated between the music and the myth, as the seam connecting and enabling their mutual relevance, he now gives music the task of mediating—a task it cannot accomplish for the very reasons Nietzsche elevates it: music allows a person to disengage from his sense of bodily particularity. While such a person can then support an idea about the rhythms of creation and destruction, he will not know himself as a kinetic image maker.

What Nietzsche, looking back on this work from 1886, sees in his early work is how his own insights into the way Attic tragedy effects an affirmation of life set him on a trajectory to oppose the conclusions about music he drew from those insights. He sees *Birth* as a stage in his own *overcoming*: his early steps in overcoming the influences of Christian morality acting on him. *Birth* was his first effort in what he comes to call his "revaluation of values": "*Revaluation of all values* [*Umwerthung aller Werthe*]: that is my formula for an act of supreme self-examination [*Selbstbessinnung*] on the part of humanity, become flesh and genius in me" (K6, 365; EH 326). In his own self-examination, then, Nietzsche foreshadows the later importance of dance for his project: as he revalues his earlier position, "dance" will emerge to figure the distance he opens between himself and Schopenhauer's pessimism, Kant's transcendentalism, and Wagner's music. Dance will represent a relationship to Christianity characterized by overcoming—a relationship in which persons learn to experience the pain they have suffered due to Christianity's ascetic teachings as a reason to love life. As Nietzsche confirms in his preface, what he has learned since *Birth*, readers can learn as well by turning to his *Zarathustra*—the book in which a dancing Zarathustra replaces the musical Socrates as the figure for a life-affirming alternative to Christian morality.

Returning Words to their Senses: *Human All Too Human*

> Life as the product of life. *However far man may extend himself with his knowledge, however objective he may appear to himself—ultimately he reaps nothing but his own biography.* (K2, 323; HH #513, 238)

Books of Nietzsche's that follow *Birth* may be read as attempts to engage the trajectory of its critique while rewriting its original conclusion. *Human All Too Human* is a case in point. What marks the distance between *Birth* and *Human* is the return of what dropped out of *Birth*: Nietzsche moves bodily becoming to the center of his work as the locus to which all ideas and actions, all philosophies and histories, must be held accountable. He formally (re)introduces "dance" as an analogy for this knowledge.

In writing *Human*, Nietzsche could not be more explicit about the distance he opens between himself and Schopenhauer on the subject of music.[14] He writes:

> In and of itself, music is not so full of meaning for our inner life, so profoundly moving, that it can claim to be a *direct* language of emotion. Rather, it is its ancient connection to poetry that has invested rhythmical movement [*die rhythmische Bewegung*], loudness and softness of tone, with so much symbolism that we now *believe* music is speaking directly *to* the inner life [*Inneren*] and that it comes *out* of it. Dramatic music is possible only when the art of music has already conquered an enormous realm of symbolic techniques through song, opera, and hundreds of attempts at tone painting . . . No music is in itself deep and full of meaning. It does not speak of the 'will' or the 'thing in itself.' Only in an age that had conquered the entire sphere of inner life for musical symbolism could the intellect entertain this idea. The intellect itself has *projected* [*hineingelegt*] this meaning [*Bedeutsamkeit*—or "meaningfulness"] into the sound. (K2, 175; HH #215, 128).

In this passage, not only does Nietzsche reject the idea that he distorted his own position to represent in *Birth*—in stating that music does not speak of "will" or have meaning "in itself"—he reconnects music to "rhythmical movement" as an enabling source. It is because music has "conquered" the symbolic techniques represented in other arts, including dance and poetry, that this "age" can even conceive of music as meaningful.

A primary task then that Nietzsche adopts in this book is to expose the physiological conditions that such a belief in music represents. He identifies a distinctively modern *decadence* characterized culturally and among individuals by forces of *desensualization* and *oversensitization*. As Nietzsche insists, an intellect can "entertain this idea" about music only when it is under the influence of a certain kind of sensory education—one in which a person learns to disregard his senses. What this education is missing, he concludes, is *dance*, meaning the kind of experiential transformation and sensory education he associates with the chorus in Attic tragedy. In short, what we see in Nietzsche's critiques of desensualization and oversensitization is how and why dance takes shape as a figure for the kind of bodily symbol-making, the extra-textual work, required by those in our "age" who seek to revalue Christian values in themselves.

Desensualizing Sense

Desensualization represents a normal and necessary trajectory in human life that Nietzsche acknowledges and even celebrates. However, his point in *Human* is that this trajectory has progressed in modern times to such a degree that it overwhelms its necessary counterweight, an education *to* the senses. On the one hand, desensualization represents the training process through which humans and cultures learn to use language, communicate, think, and believe in themselves as individual beings with consciousness and an innerliness. It is a process by which humans develop the ability to think about music in isolation from other arts and experience music as "in itself" meaningful. This trajectory is evident, for example, in the description of tragedy's birth, as Dionysian energies find expression in a bodily symbol-making whose rhythms are further constrained and stylized to catalyze new developments in other arts as well. So too in Nietzsche's critique: the fall of tragedy occurred when the trajectory of desensualization reached to such a degree that writers, actors, and spectators came to privilege the meaning of the dramatic word above rhythmic movements of the chorus and their physiologically-enlivening effects. At that point tragedies lost their ability to function as either art or religion. They served as texts, accessible to reason.

In the year following the publication of *Birth*, Nietzsche elaborated a second perspective on desensualization that his discussion in *Human* presupposes in an essay he never published, "On Truth and Falsity in their Extra-moral Sense" (1873). In this essay, although Nietzsche is tackling issues concerning the origins of language rather than art, he offers a theory of metaphor that links the two. Instead of explaining language-use as a function of reason or intellect, Nietzsche identifies the impulse to use language as a result of boredom and necessity, ignorance and forgetting. In seeking regularity people learned to make equivalences between dissimilar phenomena, taking one as a representative of the other: "Every idea originates through equating the unequal [*durch Gleichsetzen des Nicht-Gleichen*] . . . through an arbitrary omission of these individual differences" (K1, 880; OTL 179). This process begins, Nietzsche asserts, with a nerve response—a quickening of limbs. Almost immediately, a person's body transforms [*nachgeformt*] this nerve impulse into a "percept," a perception of some thing. This level of association is what Nietzsche calls a "first metaphor." It is only at a second level of metaphor making that a person "copies" the perception of some thing onto an oral sound or word designed to represent that (perception of a) thing (K1, 879; OTL 178). In this two level process, then, our bodies are not only responsible for receiving nerve impulses, they are the transition from impulse to percept, from percept to sign, from first metaphor to second metaphor. The transition from one to another is a function of how a body makes and becomes images of itself.

Three points are relevant here for an understanding of desensualization in *Human*. First of all, Nietzsche identifies metaphor-making as a "fundamental impulse [*Fundamentaltrieb*] of man." This impulse or "drive" cannot be sidestepped or reasoned away "for thereby we should reason away man himself" (K1, 887; OTL 188). Metaphor-making lies at the root of all cultural and psychological phenomena, including language and art, science and religion, and our sense of ourselves as individuals and humans. We always already "produce [metaphors] within ourselves and throw them forth with that necessity with which the spider spins" (K1, 885; OTL 186).

Second, not only are we metaphor-makers, we are so because we are bodily—not intellectual—creatures. As bodies we are essentially creative beings; and how and what we create is dependent upon the strength and scale of our senses (K1, 885; OTL 186). Based on how we exercise our senses, we create our capacity for future experience and perception; we determine whether and how we will receive nerve-impulses that we, through acts of "artistic metamorphosis," transform into percepts and then into signs and ideas.[15] In fact, every moment of our thinking, sensing, and acting is changing who we are in the world. As ongoing acts of self-creating, we do not properly exist. We are becoming—the movement of overcoming ourselves.

Third, as creators, we are constantly being tempted by products of our own making to restrict and confine the work of our senses in the world—the primary level of our metaphor-making. In the process of exercising our metaphor-making capacity, we inevitably create a "pyramidal order" of old metaphors that "now stands opposite the other perceptual world of first impressions and assumes the appearance of being the more fixed, general, known, human of the two and therefore the regulating and imperative one" (K2, 881–2; OTL 181). The temptation to dwell within this order, Nietzsche suggests, finds its apex in the rational man, the man who ignores and/or forgets the sensory origins of his ideas. Such a desensualized person lives under the illusion that his intellect is his strongest faculty and that what he can think is therefore true. Here Nietzsche contrasts the person who builds "a regular and rigid world" with the "intuitive" man who is constantly seeking a "new realm of action," of metaphor making, which he finds "in *Mythos* and more generally in *Art*" (K1, 887; OTL 188).

This account of metaphor-making as a fundamental drive of bodily humans highlights what becomes a crucial ingredient in Nietzsche's embrace of dance as a metaphor. "Dance" provides persons with a means for reawakening to the primary levels of metaphor-making—the self-creating capacity of their bodily being. Dance represents a perspective from which to expose the ignorance of our bodily creativity that desensualization perpetuates. With this insight, Nietzsche sets himself up in *Human* to perform the task he began in *Birth*: to discern what is needed to resist the forces of desensualization and generate the physiological conditions capable of finding expression in values that affirm life.

Instinct for Movement

In *Human*, Nietzsche begins his account of desensualization and language-use by asserting that the "Imitation of gesture is older than language" (K2, 176; HH #216, 129). "Imitation of gesture" refers to an instinctual kinetic response in which a person who sees the movement of another unconsciously finds herself moving too, mirroring the other's movement. As Nietzsche concedes it "goes on involuntarily even now, when the language of gesture is universally suppressed, and the educated are taught to control their muscles. The imitation of gesture is so strong that we cannot watch a face in movement without the innervation of our own face" (K2, 176; HH #216, 129). Though suppressed by systems of education, Nietzsche affirms that this ability to imitate gesture is still at work in every infant's attempt to understand his or her parents and in the efforts of adults to empathize with one another. It is the same principle, moreover, Nietzsche presumed in claiming that the elemental rhythms of singing and dancing impel spectators to identify viscerally with the chorus. Spectators feel the movement they see in themselves as their own movement, and in the process, come to understand the sensations leading the other person to make the movements that are appearing to them.

At the same time, as the passage also implies, in order to learn how to use words, humans must be taught how to mute their kinetic response—to suppress the language of gesture, to control their muscles, and to perceive and respond intellectually through the use of written and verbal signs. This process as Nietzsche describes it in *Human* thus echoes his account of a two-level process of metaphor-making. As language use develops through gesture into tone, humans become less consciously dependent upon their senses, less engaged in the primary levels of metaphor-making. As Nietzsche relates, once a kinetic response registers as a percept, that gesture may be used as a sign, a kinetic image. When this gesture is accompanied by a sound, a language of tonal signs becomes possible "in such a way that first both tone and gesture . . . were produced and later only tone" (K2, 176; HH #216, 129). In this way, humans learning to speak, read, and write train themselves *not* to receive knowledge through their bodily movements. Instead they focus their minds on the meaning of linguistic signs and presume that meaningfulness is generated by the relationships among the signs themselves. Soon such persons perceive only what they know how to think. While these persons are still engaged in metaphor-making activity, they are so at a second-level only, manipulating signs that have already been (artistically rendered) as perceptions. Metaphor-making is restricted to a linguistic realm—to the generation of increasingly refined and complex dream worlds—with names like metaphysics, theology, morality. It is this process of desensualization, of arresting muscular movement and of restraining one's kinetic instinctual responsiveness that produces people who believe in music and are capable of hearing it as meaningful apart from gestural accompaniment.

While Nietzsche does not distance himself from Schopenhauer and Wagner in his account of desensualization, he does in his evaluation of it. He denies that this development is a sign of progress. What it signifies, rather, is the success of a process of *training* by which humans learn to ignore their sensory bodily selves as having any role to play in the process of generating ideas of beauty, goodness, or truth. It signifies a process of training that convinces us to believe in ourselves as thinking, spiritual minds residing in material bodies. We learn not to experience music or other arts as sensory phenomena, and instead, ask what they mean: "our ears have become increasingly intellectual. Thus we can now endure much greater volume, much greater 'noise,' because we are much better trained than our forefathers were to listen for the *reason in it*" (K2, 177; HH #217, 129–30). Increasingly intellectual, our senses "become asensual": "Joy is transferred to the brain; the sense organs themselves become dull and weak. More and more, the symbolic replaces that which exists" (K2, 177–8; HH #217, 130). In so far as Schopenhauer and Wagner imagine that music is disembodied, abstract, a copy of will, they express the dullness and asensual character of their own senses; they express their ignorance of the most fundamental level of human artistry alive in the bodily experiencing of life.

From the perspective of a person desensualized in this way, as Nietzsche assumed most of his readers to be, dance would appear as a gesture language—a crude, ungainly, inefficient, and primitive means of accomplishing what words can do better. The process of training in which we learn to use language and hear music by itself is a process in which we lose the capacity to perceive dancing as symbolic action: "while music without explanatory dance and miming (language of gesture) is at first empty noise, long habituation to that juxtaposition of music and gesture teaches the ear an immediate understanding of tonal figures" (K2, 176; HH #216, 129). Nietzsche's thesis of desensualization, then, leads almost directly to dance as a counterpoint: dancing is poised to resist the forces that are encouraging us to dull our senses and suppress our instinct for movement. For "dance," as read through Attic tragedy, represents an alternative sensory education; it represents a transformation in which a person comes to appreciate and affirm bodily becoming as the source of knowledge and morality—of value.

Oversensitized Emotions

Before turning to dance, however, it is important to note that desensualization proceeds hand in hand with a second development that Nietzsche finds in our "age": an oversensitivity to emotions. It is a development he sees manifest in the music of Wagner. At first it may seem like a contradiction to characterize modern humans as both desensitized and oversensitive. Yet the link between the two is this: as people develop the ability to wrest their thinking and feeling free from its roots in bodily movement, from the timing and rhythms of bodily gesture, they become increasingly detached from any sense of their own emotional center. They develop the ability to move through an extraordinary range of emotions in short order. Wagner's genius, according to Nietzsche, is his ability to milk this capacity. He provides listeners with an intoxicating blend of swooning, swelling, precisely-rendered emotions. Whipping the audience into a frenzy, indulging their feelings of suffering and guilt, his music numbs people's senses, removing them to otherworldly spaces and nurturing false hopes of supernatural redemption.

The effect of listening to such music, according to Nietzsche, is to dissolve further any sense of ourselves as bodily beings. When dissolved thus, he fears, we are prey to fanaticism of all kinds.

Taken together, these two trends—desensualizing and oversensitizing—comprise what Nietzsche calls *decadence*. Trained away from our dulling and overwhelmed senses, we have lost the ability to discern for ourselves what is good for us. Unknown to ourselves—"What indeed *does* man know about himself!" (K2, 877; OTF 175)—we are unable to affirm our lives or take responsibility for ourselves as creators of value. We have lost the ability to do what Nietzsche heralded Attic tragedy for accomplishing: affirming our bodily selves as artists, god-makers, made in the image of our gods. In short, we no longer *dance*—and not just metaphorically. We no longer engage rhythmic bodily movement as a discipline for incarnating knowledge of our own bodily health. Increasingly dissatisfied with our lives, we are increasingly unable to do anything in response.

A Free Spirit

> The free spirit a relative concept. *A man is called a free spirit if he thinks otherwise than would be expected, based on his origin, environment, class, and position, or based on prevailing contemporary views. He is the exception: bound spirits are the rule.* (K2, 189; HH #225, 139–40)

The trajectory of Nietzsche's critique, of course, suggests a response: an ascending movement to counterbalance the decline. In *Human*, Nietzsche coins the term that comes to represent this movement in this book and beyond. As he writes in a preface to *Human* (written in the same year as his preface to *Birth*), "I invented, when I needed them, the 'free spirits' [*die freien Geister*]" (K2, 15; HH Preface #2, 5). In the nine sections of *Human*, the free spirit first appears in section 5, "Signs of Higher and Lower Culture," at the peak of the work, after radical critiques of philosophy, religion, and art, and prior to Nietzsche's aphorisms regarding society, women and children, the state, and "man alone with himself." It is in section 5 that Nietzsche also introduces dance as "an analogy."

Who then is a free spirit? A free spirit is one who seeks knowledge. Yet the freedom in question here, in contrast to that of Kant, is not that of a rational "I" acting over and against bodily instinct and desire. The freedom of Nietzsche's free spirit is the fruit of a transformation akin to that experienced by spectators in Attic tragedy. A free spirit is one whose relationship to his bodily being is transformed such that he is able to experience his own pain and suffering—not just suffering in general—as a stimulant to knowledge. As Nietzsche writes, "your own life takes on the value of a tool and means to knowledge. You have it in your power to merge everything you have lived through—attempts, false starts, errors, delusions, passions, your love and your hope—into your goal, with nothing left over: you are to become an inevitable chain of culture-rings" (K2, 236; HH #292, 175). With his own life as a tool and means to knowledge, a free spirit has "released himself from tradition" (K2, 190; HH #225, 140). He is able to live experimentally (K2, 18; HH Preface #4, 7), and learn in and through his experiences.

The transformation of the senses Nietzsche has in mind is one that, in the preface, he likens to the experience of convalescing from an illness. In such a convalescence, a person reawakens to her senses. Nietzsche writes: "He almost feels as if his eyes were only now open to what is *near*. He is amazed and sits motionless: where *had he been*, then? These near and nearest things, how they seem to him transformed!" (K2, 19; HH Preface, 5, 8). As her eyes open, Nietzsche explains, a free spirit is cured—not from the disease per se but from *pessimism* and for reasons akin to those of the spectator of Attic tragedy. She learns to affirm life: "You had to learn to grasp the *necessary* injustice in every For and Against; to grasp that injustice is inseparable from life, that life itself is *determined* by perspective and injustice" (9). The freedom of this free spirit, then, is no longer to need to escape from suffering or seek an external meaning for it. It is a freedom to use illness as an "instrument and fishhook of knowledge" (K2, 17; HH Preface #4, 7). It is a freedom to participate creatively in the shape of one's own becoming. The "spirit" who is free is not a transcendent mind but a self-creating, symbol-making body.

From the perspective on self and life opened by this transformation of his senses, a free spirit is increasingly capable of doing what those in the modern age cannot: discerning the enduring conditions needed for his own health. He is more aware of and less vulnerable to whatever might distract him from his path of knowledge, including his own faults and vices (K2, 288; HH #288, 173). He is capable of forgiving himself and embracing these faults and vices, again as rich sources of knowledge. As Nietzsche summarizes: "And so onwards along the path of wisdom, with a hearty tread, a hearty confidence! However you may be, be your own source of experience! Throw off your discontent about your nature; forgive yourself for your own self, for you have in it a ladder with a hundred rungs, on which you can climb to knowledge" (K2, 235–6; HH #292, 174). *The free spirit is free to do what is necessary for becoming who he is.*

The contrast Nietzsche introduces with a "bound spirit" further clarifies what is at stake for him in inventing this "relative concept." Bound spirits are bound by convention, tradition, and habit. They accept without question what is given to them as truth without needing explanation, and they come to appreciate the usefulness of this truth as proof of its veracity. The bound spirit is a Christian. However, what is bound is not the mind per se but the *senses*. Bound spirits, in training themselves to convention, tradition, and habit, live within the metaphor-edifices they have created. They train themselves to ignore their senses and in so doing restrict the field of their experience to what is familiar such that no signs of resistance even appear. In this self-reinforcing dynamic, their senses grow duller, more intellectual, and their instincts weaker. Desensed and oversensitive, they are unable to experience their own experiences as a "ladder" to knowledge. In other words, bound spirits are minds bound to a conception and use of their own body that denies bodily being any constitutive role in the production of knowledge.

As the opposite of bound, then, the freedom of the free spirit appears in greater clarity. The freedom of this spirit is not a careless play; nor an unrestrained pursuit of instinct and desire. On the contrary, the freedom Nietzsche envisions is a result of long discipline and training—but a training *to* rather than *away from* bodily becoming. Spirits who are bound feel free within the context of the idea edifice in which they live and move; they feel free from their senses and desires, free to master them, or, more frequently, to fail trying. By contrast spirits who are free are bound by their senses and desires, bound to honor and respect them as clues to the enabling conditions of their lives. They have learned to discipline themselves to their senses—to listen and attend—as the pathways by which they relate to the world, gather information, and generate knowledge. Nietzsche does not claim that a free spirit must go out and seek to live all human experiences; nor does he use his position to justify suffering in any way (K2, 339; HH #591, 249).[16] He rather insists that free spirits must find within their own experience the conditions for understanding "enormous stretches of earlier humanity" (K2, 236; HH #292, 174). As Nietzsche avers, "We may find greater value for the enrichment of knowledge by listening to the soft voice of different life situations; each brings its own views with it. Thus we acknowledge and share the life and nature of many by not treating ourselves like rigid, invariable, single individuals" (K2, 349; HH #618, 256–7).

In the end, the difference between the bound and free spirits is not one of intelligence or will, but *energy*. The free spirit has the *energy*, the vitality, needed in order to contest convention, interrogate traditions, resist habits, and develop a life-generating relation to bodily being. So important is this energy that it impels Nietzsche's driving question in section 5: "Where does the energy [*die Energie*] come from, the unbending strength [*Kraft*], the endurance, with which one person, against all tradition, endeavors to acquire a quite individual understanding of the world?" (K2, 193; HH

#230, 143). This energy is what enables a free spirit to experiment with her sensory capacity and develop alternative perspectives than those given to her from which to invite and interpret experience. Although Nietzsche admits that in his age any free spirit is a "fortunate accident," he is still interested to know how we can create the conditions that increase the chances of such accidents. He asks whether this energy is inherited or kindled; he wonders how much a person can adapt to the needs of culture and remain free.

Dance as Analogy

In Nietzsche's pondering of these questions, dancing emerges, trailing clouds of Attic tragedy, to illuminate the energy-generating discipline free spirits undertake in order to educate their senses and affirm their bodily becoming. While Nietzsche does use dancing as an image in earlier sections of the text, it is here in section 5 that he provides a formal introduction:

> *Analogy of the dance.* Today we should consider it the decisive sign of great culture if someone possesses the strength and flexibility to pursue knowledge purely and rigorously and, at other times, to give poetry, religion, and metaphysics a handicap, as it were, and appreciate their power and beauty. A position of this sort, between two such different claims, is very difficult, for science urges the absolute dominion of its method, and if this is not granted, there exists the other danger of a feeble vacillation between different impulses. Meanwhile (to open up a view to the solution of this difficulty by means of an analogy, at least) one might remember that *dancing* is not the same thing as staggering wearily back and forth between different impulses. High culture will resemble a daring dance, thus requiring, as we said, much strength and flexibility. (K2, 228–9; HH #278, 169)

A "daring dance" is what a free spirit does in creating culture. How so? In arriving at this interpretation the case of Attic tragedy is helpful. The "daring dance" Nietzsche describes between science and other disciplines bears striking resemblance to the coordination between chorus and tragic narrative in Attic tragedy. As noted earlier dance served as the connective tissue between spectator and dramatic text, enabling the spectator to identify viscerally and visually with the chorus and thus see an image of himself in relation to the drama as participating in the all-powerful rhythms of creation and destruction it represented. Given this visceral and visual identification, a spectator was able to move between the clear but tragic narrative and the intoxicating but illusory experience of choral rhythms. Similarly, in this passage, those who can accomplish a daring dance mediate between two contradictory claims in a way that enables them to affirm life. They move gracefully between the rational and the intuitive, the primary and secondary levels of metaphor-making, holding in tension the desire to dwell within a stable order and the desire to revel in the "power and beauty" of human creativity. They do so, moreover, in such a way that the success of each enterprise is ensured. Without rootedness in the senses, our knowledge would fall prey to illusions (K2, 208–9; HH #251, 153–4). Without a tension with truth, our religion, art, and metaphysics would become dead ends seducing us away from rather than toward life. Nietzsche refers elsewhere to needing a double brain (K2, 209; HH #251, 154) or to the "two heterogeneous powers" within a self (K2, 227; #276, 168).

Even though Nietzsche does not explicitly state in this passage that free spirits actually dance, his use of dance as an analogy implies that the freedom of the free spirit expresses the energy that results from a transformation of our relationship to bodily being in which we come to appreciate our embodiment as a rhythm of its own reflexive becoming—its metaphor-making activity. It suggests that the ability to move between the illusion of science's truth and the truth of art's illusion not only expresses our health but enables it. It is in the force of moving from one to the other that a given mode of valuation dislodges idea edifices and stimulates the metaphor-making capacity it presupposes. In this movement, we reclaim a relationship to our bodies as rhythms that produce ideas, both scientific and religious, as a function of, medium for, and guide to our ongoing becoming.

Still, the question lingers. The use of this analogy and the tantalizing "at least" make it inevitable: must a free spirit actually practice dancing? Although Nietzsche does not offer a definitive answer at this point in his writing life, a raft of other discussions in *Human* suggests reasons why learning to move between the realms of science and art may require practicing the kind of physical, kinetic-image-making action that Nietzsche uses "dance" to represent.

For one, in investigating the kind of energy free spirits enjoy, Nietzsche repeatedly uses images of physical vitality: a free spirit can move with a "hearty tread" (K2, 235; HH #292, 174); "step forward" when overcome with feelings of agitation, and confront the causes and not simply the affects of suffering (K2, 206; HH #248, 152); he can "leap" (K2, 225; HH #273, 166) and "his muscles move, now to the old system, now to the new" (K2, 206; HH #248, 152). Nietzsche suggests as well that physical exercise is an integral and ongoing part of a free spirit's life when he writes of the free spirit's transformation: "Better division of time and labor; gymnastic exercise become the companion of every pleasant leisure hour; increased and more rigorous contemplation, which gives cleverness and suppleness even to the body—all this will come with it" (#250, 153). In such instances, then, Nietzsche implies that a free spirit practices a responsiveness between "mind" and "body" in which the health and strength of each begets the health and strength of the other. From the perspective of a free spirit, mind and body are not perceived as opposites, and any such conception of dualism is an illusion "based on an error of reason" (K2, 23; HH #1, 13). Her actions express and reinforce an alternative perception.

Moreover, when Nietzsche uses dance imagery specifically, he does so in a way that presumes a lived experience of what it is like to dance or, of what it is like to learn how to dance. For example, in an aphorism titled *Books that teach us to dance*, Nietzsche illuminates a "feeling of high-spirited freedom" by referring to an experience: "as if man were rising up on tiptoe and simply had to dance out of inner pleasure [*innerer Lust*]" (K2, 170; HH #206, 124–5). In order to understand what Nietzsche means in using the metaphor, a person must know what it is like to experience a swell of energy coursing through the body, to direct that energy in and through particular patterns of movement, and to feel the power of being released into movement. A person must also know that the bodily movement, as dance, is *symbolic*; it engages the symbol-making capacity of the body to create kinetic images (out) of "inner pleasure" or desire. Further, Nietzsche's suggestion that books are *teaching* us to dance implies that dancing provides a critical perspective on what is written and how it is read. If we are not stirred to dance as we read—stirred to discipline ourselves to our senses in the pursuit of knowledge—then that book (or our reading of it) by not stirring our desire to dance is perpetuating the illusion of an oppositional relation between mind and body that binds a bound spirit to tradition and convention.

A second example thickens the thesis. Nietzsche writes:

> To escape boredom, man works either beyond what his usual needs require, or else he invents play, that is, work that is designed to quiet no need other than that for working in general. He who is tired of play, and has no reason to work because of new needs, is sometimes overcome by the longing for a third state that relates to play as floating does to dancing, as dancing does to walking, a blissful, peaceful state of motion: it is the artist's and the philosopher's vision of happiness. (K2, 346; HH #611, 254)

On the one hand, in this passage dance is a vision—an ideal for the artist and philosopher of happiness. It represents an alternative to either work or play, one characterized by a "blissful, peaceful state of motion [*Bewegheit*]." Yet this description is quite concrete. Understanding the description requires having had an experience of a "state of motion," and thus a physical consciousness of the differences between walking and dancing, dancing and floating. It also suggests that dancing is between walking and floating, that is, as an ideal, dancing involves a relationship to earth in which a person is neither bound nor disengaged, but engaged and free, moving between earth and sky.

Further, the ways Nietzsche uses more general images of taking steps and doing exercises also presume an experience of how the practice of dancing produces changes in one's experience of bodily being. Speaking of style in poetry, Nietzsche recounts how poets gradually learn to "get beyond realism" by limiting themselves, by learning "to step with grace, even on the small bridges that span dizzying abysses" such that "one takes as profit the greatest suppleness of movement, as everyone now alive can attest from the history of music." He continues: "Here one sees how the shackles become looser with every step until they finally can seem quite thrown off: this *seeming* is the highest result of a necessary development in art" (K2, 181; HH #221, 133). Nietzsche alludes to the analogy of dance to explain how the illusion of spontaneity and freedom, the "greatest suppleness of movement," is actually the result of self-limiting discipline. Conversely, speaking of pleasure in knowledge, Nietzsche explains how through knowledge "we gain awareness of our power [*Kraft*]—[for] the same reason that gymnastic exercises are pleasurable even without spectators" (K2, 209–10; HH #252, 154). The implication here is that the act of learning to dance gives us an awareness of our power—of our capacity for bodily becoming—that enables the "high culture" that Nietzsche uses "dance" to describe.

In these and other cases, Nietzsche's use of movement and dance imagery suggests that the process of learning how to move the body in different ways is itself a method for acquiring knowledge about one's embodiment, and thus about other humans and all of life. In learning to dance a person works with the medium through which she lives; she generates the energy, health, and physical vitality that provide her with a perspective from which to criticize approaches to morality and knowledge that isolate reason over and against embodiment. In learning to dance a person actively participates in transforming the medium through which she lives: she exercises her capacity to imitate gestures; she disciplines her attention to the effort required to recreate such gestures; she learns to stretch and expand her awareness of experience, and to appreciate such experiences as metaphors, creative and reflexive acts, and thus resources for knowing.

Nietzsche's use of dance imagery, in this regard, supports those who read him as rejecting a return to a natural or pure body. There is no such *thing*. All we can do is attune ourselves as much as possible to the edges and fissures within the edifices of our ideas where knowledge is being generated in and through our bodily becoming. It is a process that itself demands radical individuality—attention to one's particularity—for "no generally valued concept can be set up" for either opinions or health: "What one individual needs for his health will make another ill, and for more highly developed natures, many means and ways to spiritual freedom may be ways and means to bondage" (K2, 233; HH #286, 172). Rather than rely on a formula, on tradition, or on obedience to the example of others, we need to cultivate an awareness of how everything we sense and think and feel and do cooperates in bringing into reality the world in which we live. As Nietzsche writes: "If one considers, then, that a man's every action, not only his books, in some way becomes the occasion for other actions, decisions, and thoughts; that everything which is happening is inextricably tied to everything which will happen; then one understands the real *immortality*, that of movement: what once has moved others is like an insect in amber, enclosed and immortalized in the general intertwining of all that exists" (K2, 171; HH #208, 125–6).

By using dance imagery in ways that presume a lived experience of full-body symbol-making, Nietzsche pulls readers to act in ways that resist the trajectories of desensualization and oversensitization in modern culture. People whose senses are dull, weak, and overly intellectual will not understand Nietzsche's dance imagery. Yet, in so far as they have the slightest desire to understand what he uses that imagery to describe, the implication is clear: they must learn to exercise their kinetic-image-making capacity. Only then will they appreciate what "dance" as an analogy can mean as an ideal and practice for free spirits. In other words, to understand what self-knowledge entails, readers must practice exercises that will allow them to use dance as a metaphor. In making this demand, Nietzsche brilliantly enacts his theory of metaphor: he calls readers to be aware of how their knowledge of themselves and their worlds is inseparable from their bodily being in the world—from their biographies, their "temperaments" (K2, 54; HH #34, 37), the "chemistry" of their moral, religious, and aesthetic feelings and ideas (K2, 24; HH #1,14), and their "feelings of pleasure and pain" (K2, 39; HH #18, 25). Readers cannot hope to know themselves without engaging in bodily practices that train them how to participate consciously in the rhythms of their bodily becoming—their primary level of metaphor-making.

Finally, the intensity of Nietzsche's polemic against scholars in this work further supports the idea that dancing, for Nietzsche, is a kind of education needed to train the senses in an opposite and complementary direction to that opened by the acts of reading and writing. While Nietzsche admits he was a scholar, and in some respects, still is, he also offers a perspective on scholars enabled by his (analyses of) dancing. He discerns how scholarly practices of reading and writing perpetuate the dynamics of desensualization and oversensitization. As he warns: "*Caution in writing and teaching.* Whoever has once begun to write and felt the passion of writing in himself, learns from almost everything he does or experiences only what is communicable for a writer" (K2, 167–8; HH #200, 123). Because scholars feel the passion of writing they seek to maximize it, and do so by reducing the range of their experiences to those that writing can represent. As a result they cut themselves off from the primary level of metaphor-making: "though their spirit may be willing enough: their flesh is weak . . . [T]he pedantry of science and out-of-date, mindless methods have made them crippled and lifeless" (K2, 208; HH #250, 153). The spirit is strong in so far as scholars can appreciate theoretically the value of art, religion, and sensory experience; the flesh is weak in so far as scholars lack the energy required to contest conventions and engage in the rhythm of destroying and recreating their metaphor mansions. In Paul's letters, strong spirit and weak flesh signal a need to depend on Jesus for forgiveness. In Nietzsche's revaluation, strong spirit and weak flesh signal the need for bodily practices other than writing that serve to animate kinetic responsiveness, and cultivate physical consciousness of bodily becoming.

Human concludes with a vision that Nietzsche describes as a gift to the free spirit. The free spirit is walking into the mountains. "[I]n the dawning light he already sees the bands of muses dancing past him in the mist of the mountains. . . . He strolls quietly in the equilibrium of his forenoon soul" merry and contemplative (K2, 363; HH #638, 266–7). The free spirit is able to see dancing, to view it as an inspiration and as a parable for a love so great that it encompasses all of life. This vision is for the free spirit a fruit of and reward for hard work; it is a refreshing vision of happiness, beauty, and peace. It provides the free spirit with the energy needed to keep dancing himself, crossing over continually from what was to what will be.

This vision foreshadows the approach of one who, for Nietzsche, epitomizes the free spirit. Zarathustra walked in the mountains. And Zarathustra not only saw muses dancing, he was a dancer himself. Zarathustra learned how to dance, and in so doing, found the energy within himself needed to transform his experience of pain into a love so strong that it overflowed into a love for all others, and for all of life. In this text and its frame, *Gay Science*, Nietzsche pulls together the pieces he introduces here: the free spirit is one who dances.

Chapter 2

Free Spirits

*I want to learn more and more to see as beautiful what is necessary in things;
then I shall be one of those who make things beautiful.* Amor fati: *let that be my
love henceforth!*

—K3, 521; GS #276, 223

By the time Nietzsche published *Human* he was near the end of his rope.
His symptoms were preventing him from sustaining his schedule at the
University of Basel. His sister Elisabeth was no longer willing or able to keep
house for him as she had since 1875. Nor had the leave during which he had
written *Human* helped him recuperate significantly. In 1879, Nietzsche
resigned from academic life, retiring on a small pension. For the next
ten years, he lived a nomadic existence, moving with the seasons through
Italy, Switzerland, and southern France, in search of climates most conducive to his health.[1] Wherever he went he walked and wrote and walked.
He walked to open spaces of health within himself—spaces in which he
welcomed ideas and values that would guide him in realizing the *great
health* he sought for himself: an ability to overcome and revalue the life-
denying aspects of Christian values in himself, and love life. All of it. It is
a wonder he did not simply give up.

A couple of years after resigning from Basel, one such health-enabling
idea came to him while walking. It was August of 1881. Nietzsche called it
the "highest formulation of affirmation that is at all attainable": the idea of
eternal recurrence (K6, 335; EH 295). This idea provided him with the
perspective on Christian values he was seeking—an alternative to Christian
love capable of doing for his age what he perceived Attic tragedy had accomplished for the ancient Greeks: catalyzing a radical affirmation of life.

Two books followed: *Gay Science* (1882, 1886) and *Thus Spoke Zarathustra* (1883–1885). Although Nietzsche admits that the idea of eternal recurrence marks the "fundamental conception" of *Zarathustra*, he wrote the first four books of *Gay Science* first. Then he wrote four parts of *Zarathustra* before returning in 1886 to write the preface and Book Five of *Science*.[2] Not only, then, does *Science* provide a frame for *Zarathustra*, but taken together, as an introduction to and expression of Nietzsche's "highest formula of affirmation," the two books represent a peak of his career. After *Zarathustra* Nietzsche read his early books as stages in his own sickness and convalescence leading up to the high point of health that Zarathustra represents. Correlatively, in these texts, Nietzsche repeatedly refers back to *Zarathustra* as he carries out the critique of Christian values his *Zarathustra* enabled him to develop. From *Zarathustra* on, the free spirit who demonstrates the energy to which Nietzsche himself aspired is a dancer, Zarathustra.

In the dance imagery that pervades these entwined books, we find clues to how Nietzsche was able to embrace his sickness as a stimulus to knowledge, health, and ultimately to a radical love of life. In both books Nietzsche uses "dance" in ways that carry over meanings accumulated in chapter 1: dance involves visceral identification, kinetic responsiveness, gestural imitation, energized senses and instincts, and a physical consciousness of bodily becoming as a means to generating both scientific and religious ideals. At the same time, Nietzsche's references to dance pull these facets together into new shapes. Dancing appears as *therapeutic*, in helping people purify their bodies, discharge emotions, and reeducate their senses. Dancing appears alongside writing as *theopraxis*: a complementary practice for creating and becoming our highest ideals of self—ideals that do not express a "fear of *over-powerful* senses [*ubermächtigen Sinnen*]" (K3, 624; GS #372, 333), but remain faithful to the earth (Z 125). Dancing thus emerges as playing a primary role in the revaluation of values, Christian in particular.

Nietzsche's dance analogy from *Human* further provides justification for reading these works together by illuminating why Nietzsche chose to communicate his formula of radical affirmation through the contrasting styles of these two books. *Science* is a philosophical text, featuring aphorisms and prose while *Zarathustra* is a fictional account about a man named Zarathustra and his various encounters. Yet, *Science* is also book-ended with poems out front and songs in the back; it is peopled with imaginary characters, including the "mad man" who announces that God is dead. Zarathustra by contrast speaks philosophical treatises to his followers.[3] Within and between these works, in other words, there is evidence of the "daring dance" Nietzsche describes in *Human*. *Science* represents a "science" that knows its dependence on faith, art, and illusion; *Zarathustra* represents an art that knows itself as the "counterforce" to honesty, the "*good* will to appearance [*guten Willen zum Scheine*]" that makes knowledge of our error and delusion bearable (K3, 464; GS #107, 163). From this perspective, readers interested in understanding what Nietzsche means by the idea of eternal recurrence cannot restrict themselves to one or the other of these works. While *Science* does provide a philosophical expression of the idea, it is *Zarathustra* that effects a transformation that enables humans to develop the perspective, strength, and health needed to embrace the idea of eternal recurrence with gratitude as an expression of highest affirmation. In the "daring dance" of these two books, readers learn that they must themselves learn to dance not only in order to love life, but to do philosophy.

Dancing to Know: *Gay Science*

Faced with a scholarly book. *We do not belong to those who have ideas only among books, when stimulated by books. It is our habit to think outdoors— walking, leaping, climbing, dancing, preferably on lonely mountains or near the sea where even the trails become thoughtful. Our first questions about the value of a book, of a human being, or a musical composition are: Can they walk? Even more, can they dance? (K3, 614; GS #366, 322)*

In Book Three of the *Gay Science* Nietzsche first announces his most (in)famous line: "God is dead" (K3, 467; GS #108, 167). While much ink has been spilled debating this claim, Nietzsche takes it for granted. For him it means that "belief in the Christian god has become unbelievable" (K3, 573; GS #343, 279). That belief is unbelievable due to forces Nietzsche has been documenting all along—the textualization of art, the rationalization of knowledge, the desensualization of culture. While it might seem that such forces might make belief in God ever easier than before, as people are already living in ideal worlds of their own creation, Nietzsche argues the reverse: these forces of modernity promote *decadence*. They enable us to maintain the illusion that our ideals are true and represent reality. As a result, those who profess belief in God and those who deny God's existence participate in the same "killing" of "God": they treat God as an object that can be proven true or false. Nietzsche's claim is astute in this regard: he does not argue that God does or does not exist. Rather, he implies that "God" as an ideal has lost its power to affect the senses. God is dead when "God" becomes synonymous with a rationally-accessible "truth."

While Nietzsche probably suspected this death for some time, it is here, in *Science*, that he feels free to announce it. In *Science* his concern is with helping others to develop a similar ability to acknowledge and greet this insight. He is free because he has found within himself a perspective—his idea of eternal recurrence—from which to look back and embrace God's death as a reason to love life. And as detailed in the discussions that follow, "dance" appears time and again to describe the means and fruit of this "magic transformation." In Books 1 through 4, dance appears as the sign of the extra-textual work needed to counter the effects of the forces that have killed "God." In framing dance in this way, *Science* not only introduces the challenge that Zarathustra will face in overcoming himself and revaluing all values, it elucidates why Zarathustra is, from beginning to end of *Zarathustra*, a dancer.

Book One

In Book 1 Nietzsche launches a critique of Christian values—of what has been assumed in western Christian culture as good, moral, and noble. The critique is familiar to readers of *Birth* and *Human*: Nietzsche blasts Christian values for corrupting the instincts (#96), suppressing the passions (#47), weakening the senses (#48), and in general inducing an overall igno-rance of and hostility to bodily becoming. In the third to last aphorism of this section, Nietzsche introduces his contrasting position, and it features dance. He writes:

> Appearance is for me that which lives and is effective and goes so far in its self-mockery that it makes me feel that this is appearance and will-o-the-wisp and a dance of spirits and nothing more—that among all these dreamers, I, too, who "know," am dancing my dance [*meinen Tanz tanze*]; that the knower is a means for prolonging the earthly dance [*irdischen Tanz*] and thus belongs to the masters of ceremony of existence; and that the sublime consis-tency and interrelatedness of all knowledge perhaps is and will be the highest means to *preserve* the universality of dreaming and the mutual comprehen-sion of all dreamers and thus also *the continuation of the dream*. (K3, 417; GS #54, 116)

Though Nietzsche has not yet mentioned that God is dead, this account of appearance foreshadows what is to come. In it he undermines the meta-physical foundations for belief in God as a supernatural being. In this account, there is no other world than the world mediated to us through our bodily selves; all we believe and value and know is derived from our senses and shaped by our instincts. As a result, we have no access to any reality behind appearances, and all of our attempts at knowledge simply preserve the veil, the dream, that hides reality from us. Apollinian energies resonate through this vision.

At the same time, Nietzsche describes a transformation in experience that enables a person to appreciate her own participation in this realm of appearances, and thus experience the vision as a stimulus to her own creative activity. That Nietzsche calls upon dance to describe this transformation is not surprising given the analysis in chapter 1. The logic is familiar: the dance of spirits mediates a transformation in which the one who watches identifies viscerally and visually with what she sees. In doing so, she comes to experience herself differently. No longer is she a rational thinker detached from the objects she contemplates, she "too" is "dancing her dance." She knows her bodily self as participating in the eternal process of becoming and doing so through her own movement, in her own acts of metaphor making. There is a circular dynamic at work as well. It is because "I" am dancing that I am able to see the dance in the first place and feel myself dancing in the second place. Dance, then, based on how Nietzsche uses the image, stands for a practice through which a reflexive bodily knowledge is possible—a physical consciousness of myself as a bodily becoming. It is not "God" that gives meaning to me. In and through my dancing, I give meaning to myself and to my god. My dancing, and not divine will, is what *preserves* the universality of dreaming and the *continuation* of the dream—a dream that includes whatever ideals or gods I hold dear. This passage, in short, affords a first glimpse of dance as a *theopraxis*.

Book Two

This reflection on appearance in Book One provides the link to Book Two where Nietzsche moves from a discussion of metaphysics to a discussion of art as "the *good* will to appearance" (K3, 464; GS #107, 163). Of the five books in *Science*, this one contains the most references to dance, many of which refer to historical cases of dancing. Through these references, dance appears as a form of art that is the paradigm for what art should be—an art capable of providing the necessary counterbalance or transforming experience to the Book One insight that all knowledge is appearance. Nietzsche thus primes his readers to be able to receive and respond to the madman's claim. As he confirms: "If we had not welcomed the arts . . . the realization that delusion and error are conditions of human knowledge and sensation—would be unbearable. *Honesty* would lead to nausea and suicide. But now there is a counterforce against our honesty that helps us to avoid such consequences: art as the *good* will to appearance" (K3, 464; GS #107, 163).

The definition of art that undergirds this section is one that draws Nietzsche's earlier theories of metaphor-making and the origins of language closer to his emerging vision of dance. Here, humans are not only fundamentally artistic at the two levels of metaphor-making, but the metaphors they make are themselves creative. The words and ideals we generate become our bodies; they bring into being as real what they purport to represent. As Nietzsche explains, "what things *are called* is incomparably more important than what they are . . . [reputation, name, appearance] grows from generation unto generation, merely because people believe in it, until it gradually grows to be part of the thing and turns into its very body" (K3, 422; GS #58, 122). With this description Nietzsche teases out the logic already at work in earlier mentions of dance and applies it to art as a whole: bodies, in making metaphors, not only generate ideas about themselves and their worlds, but in so doing they actually create and recreate themselves. They open and foreclose avenues for further experience and response. In short because all knowledge is appearance mediated to us through our bodies, our bodies are always changing. As Nietzsche will assert in Book 3, it is an error to believe in the body, yet this error has shaped all our "higher functions, sense perception and every kind of sensation" (K3, 469; GS #110, 169).

Given this understanding of bodies and metaphors, Nietzsche accords art a privileged place among human activities. It is not just that humans are essentially artistic as noted above; but that the practice of art is important for developing knowledge of ourselves as bodily metaphor-makers. Nietzsche accords art the responsibility for teaching humans how to engage their senses in the pursuit of knowledge. With allusions to *Birth*, Nietzsche avers: "Only artists, and especially those of the theater, have given men eyes and ears to see and hear with some pleasure what each man *is* himself, experiences himself, deceives himself . . . only they have taught us the art of viewing ourselves as heroes—from a distance and, as it were, simplified and transfigured—the art of staging and watching ourselves" (K3, 433–4; GS #78, 132–3). In this conception of art, Nietzsche both lifts up theater as a model and locates the value of art in the education to sense provided by its reflexive capacity. And, as described in chapter 1, it was the dancing in Attic tragedy that enacts this reflexivity and mediates its moments such that we identify with the images we perceive ourselves as viewing and making as *ourselves*.

Nietzsche refers directly to dance as one art among others in a discussion of poetry that follows—Greek poetry. Quickly turning his attention away from the words themselves to their theatrical uses and contexts, Nietzsche revives the story of desensualization he told in *Human*. Why, he asks, did the Greeks allow rhythm to "permeate" speech" (K3, 440; GS #84, 138)? In answering this question, Nietzsche elaborates his idea in *Birth* of art as the "saving sorceress [*heilkundige Zauberin*], expert at healing" (K1, 57; BT 60). The Greeks used rhythm, Nietzsche contends, to allow listeners to discharge their emotions, purify their souls, and calm their minds (K3, 441; GS #84, 139). This discharge succeeded, however, only when listeners *danced*. Again, as in the early sections of *Birth* Nietzsche describes an aesthetic experience whose transforming efficacy depends upon a cooperation of music and dancing. As Nietzsche recounts:

> [L]ong before there were any philosophers, music was credited with the power of discharging the emotions, of purifying the soul, of easing the *ferocia animi* [ferocity of the mind]—precisely by means of rhythm. When the proper tension and harmony of the soul had been lost, one had to *dance*, following the singer's beat: that was the prescription of this therapy. That is how Terpander put an end to a riot, how Empedocles soothed a raging maniac, and how Damon purified a youth who was pining away, being in love; and this was also the cure one tried to apply to the gods when the desire for revenge has made them rabid. Above all, one sought to push the exuberance and giddiness of the emotions to the ultimate extreme, making those who were in a rage entirely mad . . . in order that the deity should feel freer and calmer afterward and leave man in peace. (K3, 440–1; GS #84, 139)

In this account, the value of the "singer's beat" does not lie in the sounds alone. The meaning of the sound is delivered only in so far as persons who listen to it dance. The point is not that people must hear the music and then decide to dance to it. The point is that dancing is itself the form in which listening takes—a listening that is visceral and embodied.

The reason why dancing is necessary to release the therapeutic value of music is implied by the foregoing discussions of art and metaphor-making: only when I hear the music in and through my moving body will the listening transform my senses and educate my instincts. If I listen passively, I respond only in and through my mind; I wonder what the music means. I generate conceptions of its meaning that inhibit me from ever encountering the music as a sensory phenomena. In short, simply listening to music preserves the illusion of a dichotomy between mind as the self and body as a real thing. Such listening thus perpetuates the causes of strife—the apparent reality of a conflicted heart, mind, and soul. In this way as well, Nietzsche provides himself with criteria for evaluating art: art that encourages passive listening fails as art or as religion. It does not make beautiful. It does not discharge the emotions. It does not transfigure experience. It simply provides an ideal world, an escape from the world, into an abstract realm of emotions. By contrast, art is effective if it stirs my kinetic responsiveness, providing me with a new experience of my own reflexive becoming. Thus, dancing, in a cooperative tension with music, emerges from this discussion as the paradigm for a kind of art that succeeds in effecting a healing change in perception. As Nietzsche intones: "What good is all the art of our works of art if we lose that higher art, the art of festivals?" (K3, 446; GS #89, 144), the art in which people listen by dancing. It is such art that provides persons with an opportunity to re-embody their thoughts and emotions, to stage and watch themselves, and educate their senses to their bodily becoming.

Soon after emphasizing the importance of dancing as a paradigm for a healing art, Nietzsche stages his own dramatic moment of self-reflection, and in the process revalues the practice of writing in light of his vision for art. "Why do you write?" he asks himself. In response (to himself), he says: "I am not one of those who think with an inky pen in their hand, much less one of those who in front of an open inkwell abandon themselves to their passions while they sit in a chair and stare at the paper. I am annoyed by and ashamed of my writing; writing is for me a pressing and embarrassing need . . . I have not discovered any other way of getting rid of [*zu losen werden*] my thoughts . . . I must" (K3, 448; GS #93, 146). His answer to himself here echoes the earlier descriptions of dancing and music as therapeutic. Writing has value for him in so far as it functions as an art, as dancing does. It does not provide him with an escape into a world of unrestrained passion; it does not express what he does while sitting or thinking. Writing allows him to discharge his thoughts and the emotions they represent. He writes down what the process of sensing, desiring, experiencing stirs in him, and does so in order to get rid of it. He writes not as a replacement for experience but in order to remain open to the gifting of experience. To write is a *necessity*, made necessary by the force of bodily living and creating.

In one sense, this self-reflection on writing simply continues the critique of textualizing knowledge that he began in *Birth*: writing is not transparent to truth, reason, or experience. Words alone have no meaning. In another sense, it heralds a new approach to writing philosophy, one he will explore in *Zarathustra*. In the wake of the death of God, Nietzsche contends, writing has meaning as an art in which our bodily selves create ideals. Writing, even philosophical writing, is as imbued with artistry, deceit, and delusion as any other human activity. However, the implication to draw is not that we should no longer write. Nor that there is no meaning or no truth. Rather, the challenge is to revive consciousness of writing as an artistic bodily practice, as a practice in and through which we exercise the self-creating potential of our bodily becoming. We must write with awareness of what it is about writing that constitutes its power.

The task is huge. People in the modern period, having lost the art of festivals, have also lost a consciousness of words as bodily-based metaphors. They no longer write as a kind of dancing, and in fact, write in ways that sustain the illusions that we are thinking minds dwelling in, over, and against sentient bodies, and that the God we can think exists. Nietzsche admits as much in describing how, in German language, the written style has come to dominate the rhythms of oral language such that the sound of the language is becoming "militarized" (K3, 462; GS#104, 161). As a result, German spirits suffer a contradiction. They harbor "a really deep craving to rise beyond, or at least look beyond, ugliness and clumsiness toward a better, lighter, more southern sunnier world" and yet suffer from "cramps": "signs that they would like to *dance*—these poor bears in whom hidden nymphs and sylvan gods are carrying on—and at times even higher deities!" (K3, 463; GS #105, 162). With this sentence Nietzsche harks back to his time at Pforta, where he was the one with cramps in whom nymphs and gods were dancing. He points forward to *Zarathustra* in which Zarathustra will claim: "now a god dances through me" (Z 153). The implication is that in *Zarathustra* Nietzsche will take up the challenge of revivifying consciousness of writing as an artistic bodily practice—as a kind of dancing.

Thus, in the trajectory of Book Two, dancing emerges as a perspective from which to criticize art, morality, scholarship, and religion that fail to honor bodily becoming as the source and end of their efficacy: *can they dance?* (K3, 614; GS #366, 322). Dancing stands for the activity in which we enact and come to know the process through which the ideals we create through our own metaphor-making become our bodies. Writing that dances, in turn, is writing that educates our senses in how to participate consciously and responsibly in the life of our bodily becoming.

Book Three

It is in Book Three that the "madman" appears to make the announcement toward which earlier discussions of appearance and art point: God is dead. The madman is distraught. He does not take his news lightly. It strikes him as tragic, and he knows that we need to find ways to respond. He soon asks: "What festivals of atonement [*Sühnfeiern*], what sacred games shall we have to invent?" (K3, 481; GS #125, 181). We need festivals in particular to help us do what the madman has not succeeded in doing: discharge our emotions and transform our experience of God's death from something horrific and terrifying into something beautiful and liberating. As the preceding discussions prepare us to understand, it is art, as Nietzsche defines it, that will help us attain the perspective on ourselves from which we can affirm God's death as an opportunity to exercise the creative potential of our bodily becoming. Festivals of atonement in which we take credit for God's death will generate in us the energy and physical consciousness we need to realize the ideal Nietzsche first introduced in *Human*—the free spirit—and generate new ideals. As Nietzsche confirms: "we need all exuberant, floating, dancing, mocking, childish, and blissful art lest we lose the *freedom over things* that our ideal demands of us" (K3, 465; GS #107, 164). This free spirit—a person unbound by convention, tradition, or habit, having the vitality and discernment needed to do what is necessary for her own health—is one who finds in the death of God an occasion to *love* her bodily becoming. Thus, it is art rendered on the model of dancing that will guide us in revaluing the ultimate Christian ideal: self-sacrificing love.

Book Four

The final book of *Science* that Nietzsche completed before writing *Zarathustra* encapsulates the challenges that Nietzsche/Zarathustra will face in working to transform the experience of God's death. For Nietzsche, the challenge in so doing will be to write as a way of *dancing his dance*, conscious of his play with appearances, his own dreaminess. He must create a work of art that effects a physiological transformation in his readers such that they can experience the idea that God is dead as a cause for immense joy—as reason to *love life*. He must write to express his own love for humanity in providing them with new ideals—the overman [*Ubermensch*] and eternal recurrence—as alternatives to the life-denying ideals provided by the crucified Jesus or the musical Socrates. He must write in ways that awaken in readers a passion for their own health, for unfolding their creative potential as metaphor makers. Thus Nietzsche will write *Zarathustra* to scrutinize experience (#319), refine senses (#302), create new tablets of value (#335), and make beautiful what is not (#299). As he insists, we "want to be the poets of our life—first of all in the smallest, most everyday matters" (K3, 538; GS #299, 240).

The task Nietzsche undertakes in *Zarathustra*, then, is not to make a rational argument about why eternal recurrence is his highest formula of affirmation; nor will he defend the possibility of creating new values.[4] Such projects would cede the terms of the conversation to the ideal Nietzsche stands to contest—the ideal of reason operating over and against experience. Rather, Nietzsche writes to change hearts. He writes not only to demonstrate how embedded in sensory experience our rational thinking is, but to stimulate readers to the physiological vitality that can perceive and reject life-denying ideals, and go on to generate life-affirming ones. As he confirms of readers: "I want to make them bolder, more persevering, simpler, gayer. I want to teach them what is understood by so few today, least of all by these preachers of pity: *to share not suffering but joy*" (K3, 568; GS #338, 271). It is in this sense that Nietzsche aims to write a tragedy. The death which readers must learn to experience anew is not Zarathustra's but God's.[5]

Nietzsche approaches this task by enacting his ideal of such a transformation as it occurs in a fictional person: "I want to create for myself a sun of my own" (K3, 551; GS #320, 254). Zarathustra is his sun. At the beginning of *Zarathustra*, Zarathustra already knows that "God is dead." In the course of the book's four parts, he learns what he thinks he already knows: how to respond. That Zarathustra is a dancer, for reasons now evident, signals this process: he is participating in the process of overcoming the (Christian) ideals and values that have become his body. Zarathustra is dancing his dance, living in constant attunement to the ever-changing patterns of sense, habit, pleasure, and pain that he at any given moment is. For this reason, as readers come to learn, Zarathustra is able to transform his experience of his "most abysmal thought" (Z 327)—eternal recurrence—and affirm it as the highest expression of love for life.

Zarathustra's Lesson

Zarathustra is all about love. Love is what Zarathustra has discovered for himself as the story begins. Love is what impels him to leave his solitude and teach humans. Love is what he himself learns to do through his adventures. The love he has and shares and learns is a love for his own self as a process of its own becoming—a love so strong that it rages against all of the ways in which humans weaken themselves, settling for less of the joy, the energy, the life than they are capable.

In the reading of *Zarathustra* that unfolds in this chapter, Nietzsche's images of *dancing* provide a key to interpreting the meaning and efficacy of the text. Nietzsche's images of dancing occasion a *visceral identification* between reader and Zarathustra such that readers come to *hear* Nietzsche's rhythms of poetry and song *through their bodies*. They feel compelled to *imitate* Zarathustra's *gestures*; *discharge emotions* of fear and bewilderment; *educate* their *senses* to the rhythms of bodily becoming, and engage in their own acts of *theopraxis*, making and becoming their highest ideals. Nietzsche's dance images in this text, in short, appear as revaluing the model of love represented in the Christian incarnation. As dancing, love is not an act in which God becomes human and sacrifices Himself in order to redeem humans. Love is an act in which human bodies create and become images of their highest aspirations and in doing so redeem themselves. What readers will learn from Zarathustra's *dancing* is how to love themselves with a "holy selfishness" so strong it overflows to encompass all of life.

Walking a Tightrope

The link between Zarathustra's love and the dance images that reveal it surfaces in the first pages of the First Part. As the Prologue begins we learn that Zarathustra has spent ten years alone in the mountains with his eagle and his serpent and that he is "weary of his wisdom." He yearns for the opportunity to share it: "I would give away and distribute, until the wise among men find joy once again in their folly, and the poor in their riches" (K4, 11; Z 122). The allusions to Jesus and the Twenty-third Psalm ("My cup overfloweth") prefigure what we are soon to learn: what must overflow from Zarathustra's cup is love—not a Christian but a Dionysian love.

That Zarathustra's love is not the love of God but of humanity emerges in the second section of the Prologue, when a hermit saint Zarathustra passes by in the forest recognizes him: "Does he not walk like a dancer?" (K4, 12; Z 123). Zarathustra responds: "I love man" (K4, 13; Z 123). This is Zarathustra's wisdom, the overflowing substance he longs to share. It is the secret of his dance-like walk. As we learn from the saint, Zarathustra carried his ashes to the mountain ten years earlier—the ashes of his faith in God, in himself, in life. And now he returns, having rekindled his flame into a love so great that it demands, for its own enjoyment, that it be shared. The hermit saint, by contrast, sings, cries, laughs and hums, but he does not dance: "I love God; man I love not. Man is for me too imperfect a thing. Love of man would kill me" (K4, 13; Z 123). Believing in an absolute difference between the perfect God and imperfect human, the saint lives out Feuerbach's claim: because the saint loves God, his image of perfection, he is unable to love himself or other humans who seem to him to fall short of that ideal. Yet Zarathustra loves even the saint so much that he continues on his way without contesting the hermit's wisdom. "Could it be possible?" he whispers to himself. "This old saint in the forest has not yet heard anything of this, that *God is dead*!" (K4, 14; Z 124).

Asking this question Nietzsche knits a nexus of images that runs through the book: *to dance is to love life*. That Zarathustra walks like a dancer is a sign that he has kindled the ashes of his faith in God into a fierce love for humans: he has *overcome* his attachment to God and *revalued* his faith. That he walks like a dancer signals his magic transformation—a transformation of his sensory engagement with the world, his connection to the earth. He now has the energy and strength to *affirm life*, to have faith in the earth and faith in the overman [*Ubermensch*] as his ideal, as the meaning of the earth. Expounding on his ideal, Zarathustra contends: "All beings so far have created something beyond themselves" and so will human beings (K4, 14; Z 124). The life Zarathustra loves in himself and in others *is* this process of overcoming oneself, for "what can be loved in man is that he is an *overture*

[*Ubergang*] and a *going under* [*Untergang*]" (K4, 17; Z 127), that is, a *dancer*.

The image of Zarathustra's dance-like walk provides the link between Zarathustra's professions of love and his experience in the next scene in a way that further identifies dancing with the human process of self-over-coming. Coming out of the forest, Zarathustra enters the market place, where a crowd has gathered to watch a tightrope walker. It is a scene to which commentators refer as a key in interpreting *Zarathustra*.[6] Few note, however, that in German, tightrope walker is "rope dancer."[7] In addressing the crowd, Zarathustra draws the connection between himself and the rope dancer: "Man is a rope, tied between beast and overman—a rope over an abyss [*Abgrund*]" (K4, 16; Z 126). We already know that Zarathustra is a dancer and that the overman is his ideal. In the rope dancer, then, Zarathustra he sees an image of his ideal, an image of himself in pursuit of that ideal, and an image of the possibility within each human that forms the object of his love. While Zarathustra has just described himself as "going under" in making the decision to share his love for humans with them, here he clarifies: to want to love humans as overmen as he does is itself to engage in a journey of self-overcoming. Readers are warned: as Zarathustra learns to share his love with other humans, he will enact for us the self-overcoming that characterizes the movement from animal to overman. He will need to overcome himself in order to realize his ideal of loving humans.

After making this connection between himself and the rope dancer, Zarathustra expounds upon the love he has to share. "I love those who do not know how to live," he intones, "except by going under, for they are those who cross over," dancing across the rope to the overman ideal (K4, 17; Z 127). Zarathustra's list of loves overflows for a formidable two pages, articulating an array of different dynamics by which humans overcome themselves—through sacrificing themselves for the earth; living to know; working to build a house for the overman; loving their virtue; being the spirit of virtue; making virtue their "addiction"; chastening their god "because he loves his god" (K4, 17–8; Z 127–8). In his penultimate profession, Zarathustra harks back to *Human*: "I love him who has a free spirit and a free heart: thus his head is only the entrails of his heart, but his heart drives him to go under" (K4, 18; Z 128). Here Zarathustra alludes to the transformation that Nietzsche/Zarathustra wants to occasion in those who read and hear: a revaluation of head and heart. A free spirit is one who embraces the overman as an ideal—one with the energy to resist dwelling in the metaphor-edifices that corral his thinking; to overcome the Christian morals in himself, and to learn to love.

Responding to the bewilderment of the crowd, Zarathustra expands the image of dancing into a criteria for the kinds of ideals he envisions, ideals that remain faithful to the earth. He states: "I say unto you: one must still have chaos in oneself to be able to give birth to a dancing star. I say unto you: you still have chaos in yourselves" (K4, 19; Z 129). The chaos within is the confusion, the sparring instincts and desires that resist easy accommodation. Out of this passionate mess, Zarathustra insists, our virtues and ideals grow. To give birth to a dancing star, then, is to create ideals for oneself—stars—out of one's chaos, one's highest, strongest desires. And not only stars, but stars that *dance*—ideals that make holy the kind of transformation of heart and head that Zarathustra himself has experienced in creating his own dancing star, the overman. Zarathustra admits, however, that this journey is dangerous: for lurking on the path is the "last man" who will ask "What is love?" and blink (K4, 19; Z 129).

What happens next in this scene is crucial. The rope dancer begins his performance. A jester appears and hops over him. The rope dancer falls. In response commentators ask: does the fall of the rope dancer foretell of Zarathustra's own failure to realize his ideal, the dancing star to whom he has given birth? Is Nietzsche communicating to us the message that suffering, death, and nihilism are inevitable? Is he signaling the impossibility of creating new values?

Reading the dance imagery in this scene in light of Nietzsche's earlier work suggests the opposite. The rope dancer does not fall *because* the jester leapt over him. The rope dancer falls because he "lost his head" when he saw his "rival win"; he stopped dancing (K4, 21; Z 131). What is important in the encounter, moreover, is not what happens but how Zarathustra experiences it. He does so through his transfiguring love. When the rest of the crowd abandons the dying man, Zarathustra kneels by his side. When the man laments that "the devil" will "drag me to hell," Zarathustra defuses his fear of punishment: "all that of which you speak does not exist: there is no devil and no hell. Your soul will be dead even before your body" (K4, 22; Z 132). When the man responds, then "I lose nothing when I lose my life. I am not much more than a beast that has been taught to dance by blows and a few meager morsels" (132), Zarathustra defuses his fear of meaninglessness: "By no means . . . You have made danger your vocation; there is nothing contemptible in that" (K4, 22; Z 132). In his responses to the dying man we see Zarathustra's love in action. In Zarathustra's hands, the death of God is a healing, comforting ideal; and it becomes so when the dying man is able to appreciate his own *dance* not as the movements of a trained beast, but as the vocation of one who creates over himself.

Zarathustra's encounter with the dying man highlights, demonstrates, and enables him to grasp his own gift. Zarathustra learns from his success in this situation that his vocation is to teach men like this rope dancer—men who make danger their vocation, who want to learn to dance across the rope, overcome themselves, and give birth to (themselves as) dancing stars. While the crowd was not amused by Zarathustra's professions of love, the dying man responded with gratitude. Zarathustra also learns what it is he must teach such rope dancers. He vows to teach them how to *dance*: he shall show them "all the steps [*Treppen*] to the overman" (K4, 26; Z 136). Flush with the joy of this new truth, Zarathustra *revalues* the tragic situation he has just endured: "To my goal I will go—on my way; over those who hesitate and lag behind I shall leap" (K4, 27; Z 136). In this way he expresses his gratitude for the jester's leap: it has helped him learn with whom and how to share his love for humanity.

The rest of the First Part, "Zarathustra's Speeches," when read in the wake of this interpretation of the Prologue, appears as Zarathustra's first attempts to teach fellow rope dancers how not to lose their heads, even when they see their rivals win. These Speeches are the most philosophical parts of the book, almost as if lifted from one of Nietzsche's works of prose. In them Nietzsche rejects the teachings of his own rivals—those who preach faith in an other world, or after world. In so far as people believe in such ideals, Zarathustra contends, they actively cultivate a hatred for all things earthly—for the body, sexuality, and art, the senses, instincts, and desires. In contrast, Zarathustra counsels love, a love symbolized by the act of dancing.

Two examples suffice. First, in "On the Despisers of the Body," the title primes the audience: Zarathustra will explain what it means to love the body rather than despise it. As he relates, making this change of heart (from hating to loving) does not involve simply replacing a low value accorded to the body with a higher one. Rather, to love the body is to experience it differently—to appreciate the body as "a great reason"; as all that "I" am; as a "doing" for whom sense and spirit, any conception of ego or I, represent "toys." For Zarathustra, one who loves the body knows it as a "self" that stands behind all "thoughts and feelings" (K4, 40; Z 146); as a movement whose "fervent wish" is "to create beyond itself [*über sich hinaus zu schaffen*]" (K4, 40; Z 147). In this speech, then, not only does Zarathustra upturn the traditional body/soul hierarchy, he unmasks both the conception and experience of such a hierarchy as products of a body's own creative action. He unmasks the dichotomy as an expression of a weakened body—one that has lost the ability to create beyond itself. It is only when a body/self "wants to die and turns away from life," Zarathustra insists, that it creates images of itself as a thinking "I" who exists to deny the instincts and numb the senses of a material object called "the body." Only a sick body creates an ideal of itself as an inherently sick thing, meriting rejection. Yet as Zarathustra is aware, "I" and "body" are metaphors that have become us. They represent a last ditch effort by the body/self to create something rather than nothing; such ideals produce bound spirits, weak and impoverished lives.

Thus when Zarathustra urges his listeners to have faith in the "body and its great reason," he is calling them not to think differently about the body but to transform their experience of being body (K4, 39; Z 146).[8] He is calling listeners to *love* their bodies *because* their bodies are imperfect and because they are *processes* of creating beyond themselves. Such bodies— imperfect and becoming—are the source and substance of human's highest potential. To love such bodies, in turn, is to engage in the kinds of "doing" that will give birth to dancing stars—to ideals of oneself as a process of creative overcoming. In short, to have faith in the body as Zarathustra urges is to commit oneself to learning how to heed one's body as a great reason, to discipline one's thinking and feeling to a body as an ever-evolving wisdom.[9]

Many of Zarathustra's speeches may be read as explaining just how to cultivate such faith and love. A person must listen to the voice of her body and find in her passions the sources of her virtues (K4, 42; Z 148). She must flee into solitude (K4, 65; Z 163); develop patience and learn to wait (K4, 56–7; Z 158); obey the wisdom of her love for life (K4, 59; Z 160); seek the innocence of the senses (K4, 66; Z 166); and create by esteeming (*"Schätzen ist Schaffen"*) (K4, 75; Z 171). She must create gods for herself out of her passions (K4, 82; Z 177). In general, she must cultivate a physical consciousness—an awareness of bodily becoming—that finds expression in ideals, dancing stars, that nourish her ongoing participation in acts of primary level metaphor-making.

In "On Reading and Writing," three speeches later, Zarathustra invokes dancing over and against reading and writing as an image of the practice that better encourages love for bodily becoming. After professing his love for life (again), he continues: "I could only believe in a god who could dance" (K4, 49; Z 153). The reverse is implied: Zarathustra could not believe in a god who appears fixed in the form of an eternal Word. His ideal must dance—it must be dynamic, bodily, overcoming itself. Zarathustra does not say that he *does* believe in a god who dances. His point is that he *could*—that to believe in such a god would express his own bodily creativity. In other words, a *god* who *dances* is Zarathustra's ideal ideal—one that does not pretend to represent an eternal, other reality, but one whose meaning derives from the bodies it becomes. As Nietzsche's discussions of metaphor insist, not to make ideals is not an option; human beings are ideal-makers, god-makers, instinctively, and they *are* because they are bodies. The question is rather how to make gods that dance, ideals that serve to promote health—meaning the energy and vitality required to create and love beyond oneself. Faith in a god who dances meets this criterion in so far as it encourages appreciation of bodily becoming as the medium in and through which I can become godlike. Said otherwise, believing in a god who could dance would encourage me to remember that my ideals, my gods, are expressions of my body/self creating beyond myself. [10]

Yet the difference Zarathustra invokes here is not just between dance as an image of lightness and gravity as an image of weight: that the ideal dances also represents a relationship between Zarathustra and his ideal in which he becomes that ideal, or that ideal becomes him. As he relates in concluding this speech, he has learned to walk, run, and then fly: "Now I am light, now I fly, now I see myself beneath myself, now a god dances through me" (K4, 50; Z 153). As the dancing god dances through him Zarathustra has a double experience of himself akin to that occasioned by Attic tragedy and characteristic of the dramatic, dancing art Nietzsche describes in *Science*. Two contrasting dynamics are held together in the movement of dance: Zarathustra's dancing is a god dancing through him. Zarathustra is both subject and object—both the body making images and the body's image of itself. He is not simply both, however, but the movement from one to the other. As he identifies viscerally with his (image of) god, he finds in himself the strength and awareness needed to make the movements of his god. As he does so, his bodily being changes. In this change, he *knows* his bodily being as the movement through which he becomes who he is—a creator of values. In other words, the "dancing" in this passage signals a mode of knowing oneself in which a person knows who she is in the process of realizing ideals of who she desires to be, her dancing star. She knows herself as the movement of incarnation—god becoming human. Reading and writing, this speech implies, cannot produce the same effects.

In the three moments that comprise the final speech of the First Part, "On the Gift-Giving Virtue," Zarathustra rehearses what is necessary for the love of bodily becoming he has just taught. He calls his "brothers" to watch for moments in which their spirit—as one product of the body—wants to speak in "parables": "There your body is elevated and resurrected; with its rapture it delights the spirit so that it turns creator and esteemer and lover and benefactor of all things" (K4, 99; Z 187–8). What the spirit speaks (as does Jesus) are *parables*: their meaning does not lie in their conceptual content alone, but in the ways in which they participate in the "resurrection" of the body. One who has faith in the body appreciates such resurrection not as an act in which God raises a human body to heaven; but as an action in which a body exercises its own creative, ecstatic potency in making and becoming an ideal of itself as "spirit."

In the second moment Zarathustra's tone changes; he issues a warning. "Remain faithful to the earth," he calls, "with the power of your virtue" (K4, 99; Z 188). He counsels his followers not to let their creative idealizing, as powerful as it is, fly away into abstractions; they should rather generate ideals that encourage the listening he describes in the first moment. He urges them to create for themselves ideals that lead "back to the body, back to life, that it may give the earth a meaning, a human meaning" (K4, 100; Z 188). Gods that dance. Such ideal-creating, he admits, is both necessary and difficult for "much ignorance and error have become body within us" (K4, 100; Z 189). Because we have generated ideals that oppose and deny sensory life, we have become the pale, sick images of those ideals. As a result our ability to create ideals that remain faithful to the earth is in jeopardy. We have lost the "knowledge" that purifies and elevates the body—physical consciousness of ourselves as creators.

In the third moment of his goodbye speech, Zarathustra confirms that the path to such knowledge is one each person must find him or herself in relation to his or her own body. He counsels his brothers to go away from him, to resist him. Only if they flee into their solitude will they be able to discover for themselves the value of what he is saying. Only then will they learn to love themselves—not because he tells them to, but as an expression of their own self-overcoming. For as he had insisted earlier: "You cannot endure yourselves and do not love yourselves enough" (K4, 77; Z 173). Zarathustra has hope that they will learn.

In sum, the three moments of this final speech recap the logic of Nietzsche's dance imagery and suggest its significance in his revaluation of values: humans must awaken their sense of themselves as creators (dancers), create images that remain faithful to their bodily becoming (dancing stars), and do so by developing consciousness of their own bodily becoming (dancing). In describing the effects of so doing, Zarathustra frames this process as a revaluation of Christian incarnation: it is the people engaged in this practice who will *resurrect* their bodies and *redeem* themselves. Zarathustra's "gift-giving virtue" is a love that is not focused on some other world at the expense of earthy life, but one that encompasses a bodily existence, including the ideal-forming capacity that leads us to generate ideals of other worlds. It is a love expressed in the ideal of the "overman"—a love for what humans, because they are bodies, are always already becoming. It is a love that is only possible in the wake of God's death. For when God is dead, so too are the reasons for despising and denying our self-creating bodily becoming.

Rhythms of Love

Amidst Zarathustra's exhilarating rivers of words, the First Part hints at a small crack in his love that ruptures into a chasm in the Second Part of the book. It is an abyss Zarathustra must himself dance across, as he does in the Third Part, in order to approach his goal in loving humanity in the Fourth Part.

That crack in the First Part appears in two of the only dialogues Zarathustra has in the course of his Speeches. In one he steers a youth away from the nihilism that grips him, saying "do not throw away your love and hope" (K4, 53; Z 156). In another he reports on a conversation he had with an old woman and notes that she said: "You are going to women? Do not forget the whip!" (K4, 86; Z 179). A warning sounds in each of these encounters: there will be people whom Zarathustra will have trouble moving—he will have difficulty teaching such people to make the right steps. As we learned from the exchange of the rope dancer, a whip will work to make a beast dance, but not a human. The old woman's mocking tone suggests as much. Her response conveys a sense that Zarathustra does not yet know all that he needs to know in order to love life fully.[11] Zarathustra is teaching what he needs to learn; he must learn what he thinks he already knows.

These cracks break open early in the Second Part in the context of three songs that identify dance with a love that redeems. After learning that "Months and years passed" (K4, 105; Z 195) after Zarathustra had returned to the mountain cave, we learn that he again grew restless. He was seized with a longing for his friends. He yearned to return to them, fearful for what had happened to them. As he vividly describes at a later moment, "Verily, many among them once lifted their legs like dancers, cheered by the laughter in my wisdom; then they thought better of it. Just now I saw one groveling—crawling back to the cross" (291). The Second Part begins, then, with Zarathustra revisiting themes from the First Part, urging his friends to have faith in the earth and love their bodies. He urges them to awaken their "slack and sleeping senses" (K4, 120; Z 205) and think through them (K4, 110; Z 198).

Three songs interrupt his teachings. Zarathustra grows reflective and sings. That he sings is significant: up until this point he has spoken. As a dancer who sings, he evokes the chorus; he poses as an anonymous member of humanity. That Zarathustra is singing also implies, given the discussion in *Science*, that in order to understand him we must listen through our dancing bodies. It is in these songs that Zarathustra tells of a rift in his love for humanity akin to the one Nietzsche describes in *Birth* as a primal contradiction.

In "Night Song" Zarathustra hymns this cleft as the difference between giving and receiving. Zarathustra's ideal of the overman stirs his overflowing love for humanity; he yearns to give. Yet he is overflowing so fast that he cannot receive love in return (K4, 136; Z 218). His love impels him to seek the company of friends, to care about each one of them, to attach his heart to them. Yet, so attached, he finds it difficult to teach them that they can and must overcome themselves and learn to love. As such, despite his happiness in giving, he does not know the happiness of those who receive.

He is lonely, and out of this loneliness grows an anger toward those he loves. He is angry with them for not loving him, for not being strong enough to do so (K4, 137; Z 218). That Zarathustra feels this pain and that his feeling is a catalyst for self-knowing is what renders this song potent. The rhythms of his singing draw readers to identify with his suffering, to attach their hearts to him, imitate his gestures, and learn from his learning.

Those gestures expand in the second song, "The Dancing Song," in which the theme of the rift is amplified from personal loneliness to a Dionysian range. The song occurs in a scene where Zarathustra, seeking a well ("Life is a well of joy" (Z 210)), finds a green meadow where girls are dancing. When the girls recognize him, they stop dancing. Urging them to continue, he promises to sing along. That he watches with joy and appreciation and yearns to participate is itself a sign that he has learned, at least in part, what he teaches: his senses are innocent. He too celebrates passion, bodily movement, and Eros, the god with whom they dance. As a free spirit, it refreshes his heart to see their dancing (K4, 139–40; Z 219–20).

Yet in the song that he sings to accompany the dancing, Zarathustra describes a seemingly unbridgeable chasm in the relationship between himself and "life." Life, appearing in the form of a woman, is unfathomable.[12] And the more she tells him so, flatly in words, the less he understands. His "wisdom," another female he also loves, reminds him of life, and gives him images of life, but she does not bring him any closer to life. Life still appears as a mysterious fisher of men into whose eyes Zarathustra finds himself sinking. In his song, then, Zarathustra's love for life expresses itself in a hatred of life for being so elusive. "Deeply I love life—and verily, most of all when I hate life" (K4, 140; Z 221). His hatred is the form in which he feels his deep love for life, just as his anger in the first song is the form in which he feels his love for his friends. The rift he suffers is internal to his love itself. It is his love that begets anger and hate.

When the girls stop dancing and leave, Zarathustra is sad. Watching their dancing did not transform his experience of life. Even so, the experience of singing to their dancing did bring him to the brink of realizing what is hindering his ability to connect with his friends, love life, and dance along with the girls: "something unknown is around me and looks thoughtful" (K4, 141; Z 221).[13]

That "something" takes shape in the third song, the "Tomb Song," where readers learn why Zarathustra, the dancer, is not dancing: he has not learned to love himself enough. In this song Zarathustra remembers the errors that have become his body; painful experiences from his past are still alive within his flesh. The tombs of which he sings are the tombs of his own youth. He recalls the "glances of love, you divine moments" that died (K4, 142; Z 222). He sings of playmates, blessed spirits, and visions of hope; he remembers his vow to see the holy in all things (K4, 143; Z 223). Yet in an indictment of Nietzsche's own childhood more severe than that found in the recently discovered pages from *Ecce Homo*, Zarathustra addresses his enemies, saying: "you changed those near and nearest to me into putrid boils. Alas, where did my noblest vow flee then? . . . And when I did what was hardest for me and celebrated the triumph of my overcomings, then you made those who loved me scream that I was hurting them" (K4, 144; Z 223–4).[14] Malice hit his heart (K4, 143; Z 223). He was unable to fulfill his vow in seeing the divine in all life. Summarizing the impact of these experiences, Zarathustra sings:

[O]nce I wanted to dance as I have never danced before: over all the heavens I wanted to dance. Then you persuaded my dearest singer. And he struck up a horrible dismal tune; alas he tooted in my ears like a gloomy horn . . . I stood ready for the best dance, when you murdered my ecstasy with your sounds. Only in the dance do I know how to tell the parable of the highest things: and now my highest parable remained unspoken in my limbs. My highest hope remained unspoken and unredeemed. And all the visions and consolations of my youth died. (K4, 144: Z 224)

Nevertheless, in this song, Zarathustra claims that he did not fall into the pit of despair: his will "would walk its way on my feet" (K4, 145; Z 224). He hails his will as the "shatterer of all tombs," facilitator of all resurrections (K4, 147; Z 225). "Will to power" is "the unexhausted procreative will of life" (226), the creative becoming of his bodily being. For, he adds, "Only where there is life is there also will: not will to life but—thus I teach you— will to power," a will that finds expression in a person's creating beyond herself (K4, 149; Z 227).

Even so, in hearing the closing optimism of this "Tomb Song," readers also know that in the previous song, Zarathustra did not dance, and that his relationship to life remains troubled. The juxtaposition suggests that there are still tombs Zarathustra has not yet shattered—places where he is vulnerable. And as the "Tomb Song" suggests, those places are in his love for his friends and for life. His crushed dreams and lost loves—his inability to dance as he wanted to dance—are "secret sores" (K4, 261; Z 320) to which his parasites ("*Schmarotzer*") attach.

In the course of the Second Part, these parasites soon appear, first in the form of a Soothsayer ("*Wahrsager*") who predicts that Zarathustra will fail to transform people's experience of God's death and fall: "The best grew weary of their works. A doctrine appeared, accompanied by a faith: " 'All is empty, all is the same, all has been!' " (K4, 172; Z 245). Harking back to Zarathustra's "Tomb Song" the soothsayer adds: "Verily, we have become too weary even to die. We are still waking and living on—in tombs" (K4, 172; Z 245). When Zarathustra hears these words, they "touched his heart and changed him" (K4, 172; Z 246). He grieves for a biblically-resonant three days, and falls into a deep sleep. When he awakes, he tells his "disciples" about a dream in which tombs awakened, coffins ruptured, and horrendous grimaces, mocking laughter, and screams swirled all around him. As those who have heard his "Tomb Song," we have a sense of what these tombs might contain: the dead glances of his youth. Unrealized visions of girls dancing. As his own cry awakens him, he "came to his senses" but he still does not understand the meaning of this dream (K4, 174; Z 247).

In the speeches that follow, readers watch as Zarathustra gradually comes to understand the meaning of his own dream. He must stage his own redemption, resurrect his own body from its tombs, and learn to love himself enough. As he explains: "The now and the past on earth—alas, my friends, that is what *I* find most unbearable; and I should not know how to live if I were not also a seer of that which must come. A seer, a willer, a creator, a future himself, and a bridge to the future—and alas, also, as its were, a cripple at this bridge: all this is Zarathustra" (K4, 179; Z 250–1). Standing at the bridge Zarathustra can see the task before him. He must somehow overcome his ambivalence—his love-impelled anger and hate. He must do so by transforming his experience of that ambivalence such that he can see and love it as necessary. His creative will, as he confesses and the Soothsayer predicted, is "still a prisoner" (K4, 179; Z 251), a cripple, who cannot dance. In order to dance he must learn to *will backward*. Thus he explains: "To redeem those who lived in the past and to recreate all 'it was' into a 'thus I willed it'—that alone should I call redemption" (K4, 179; Z 251). After this speech, Zarathustra's own reaction suggests its relevance to him: he looked "like one who has received a severe shock" (K4, 181; Z 253).

In the final sections of the Second Part, Zarathustra pulls into himself and again acknowledges his ambivalence—the rift between himself and those he loves. He complains that his words have not reached humans: "I have gone to men, but as yet I have not yet arrived" (K4, 189; Z 259). At this time, the "stillest hour" speaks to him without a sound: "You must yet become as a child and without shame. The pride of youth is still upon you; you have become young late; but whoever would become as a child must overcome his youth too" (K4, 189; Z 259). Again, the message is that Zarathustra has not fully resurrected his own body. Because his will is locked in the tombs of his youth, he is not yet able to dance. He is crippled at the gate by his pride, that is, by his inability to affirm his lost loves. He responds to this soundless voice "I do not want to" and is surrounded by gales of laughter (K4, 189; Z 259).

Despite his reluctance, Zarathustra *is* a dancer, and he knows what he must do. He must follow his own advice—the advice he has given to followers and friends. He must flee into solitude, cultivate patience, listen to the great reason of his body, and learn how to redeem his own past—not by reconciling himself to what happened, but by transforming his experience of it so that he is capable of finding in it a stimulus to love. In short, to will backward he must be able to accept as his truth the "abysmal thought [*abgründligher Gedanke*]" (K4, 270; Z 327) that everything in life recurs eternally—his love, his loneliness, his anger at his friends, his hatred of life, and the jesters and soothsayers who predict his fall. Only then will he dissolve the fears that prevent him from loving his friends in ways that will help them overcome themselves and learn to love him. Only then will he free himself from the memories of failure that hinder him from dancing with life. So Zarathustra "went away alone and left his friends" (K4, 190; Z 259).

Encountering the Abysmal Thought

The Third Part of *Zarathustra* is the story of Zarathustra's journey back to his mountain cave where he finally succeeds in redeeming his past. En route, that past rears its ugly head in various forms—most notably a jester-like dwarf—who predicts again that Zarathustra's message of self-overcoming love is doomed to fail. Zarathustra resists, wielding his own wisdom in response. But it is not until he reaches his mountain peak that he summons his "most abysmal thought." We never hear it. We only see the effects of its eruption. Zarathustra falls down "as one dead" and remains lying there for seven days. During this span of time in which God created the world, Zarathustra recreates himself. At the end, he regains his senses and picks up a rose apple. His animals greet him: "the world awaits you like a garden" (K4, 271–2; Z 328).

Evidence of Zarathustra's magic transformation appears in the first words he speaks. He speaks from a Dionysian perspective, describing words and sounds as "rainbows and illusive bridges between things which are eternally apart": "Precisely between what is most similar, illusion lies most beautifully; for the smallest cleft is the hardest to bridge" (K4, 272; Z 329). This smallest cleft, between giving and receiving, body and soul, self and other, Wisdom and Life, the moments in a metaphor—all trace the cleft in Zarathustra's own heart—a cleft over which words form an *illusive* bridge. Words, he continues, merely *dance* between things: "Speaking is a beautiful folly: with that man dances over all things. How lovely is all talking, and all the deception of sounds! With sounds our love dances on many-hued rainbows" (K4, 272; Z 329). Zarathustra's dance imagery here seems almost nostalgic, as if he is longing for a bridge that does not exist and can never be. Perhaps it is the bridge to the overman. The *dancing* of words and love is what enables them to conceal the clefts that words and love can never mediate. Perhaps Zarathustra has not and cannot overcome himself.

His animals, however, respond in a way that pushes Zarathustra to a further affirmation. They hark back to the passage on appearance in *Human*: "to those who think as we do, all things themselves are dancing: they come and offer their hands and laugh and flee—and come back. Everything goes, everything comes back; eternally rolls the wheel of being . . . In every Now, being begins; round every Here rolls the sphere There. The center is everywhere. Bent is the path of eternity" (K4, 272–3; Z 329–30). That words and love dance, according to the animals, is not a problem. The "things" or souls that words and love fail to mediate themselves participate in the ongoing rhythms of bodily becoming. They too are moving images. Even Zarathustra who knows is dancing his dance. Even the image of a cleft is an illusion in whose life we participate.

Zarathustra admits that the animals are right. He recounts the horror of his seven days: "that monster crawled down my throat and suffocated me. But I bit off its head and spewed it out" (K4, 273; Z 330). The monster took the form of a "great disgust with man" (K4, 273; Z 331)—the greatest threat to his love and itself an expression of that love. It was a disgust that had power over Zarathustra due to his own secret sores, biting him in a place that cut off his ability to speak, breathe, eat and drink, or engage in the rhythms that sustain life and love.

As the scene progresses, readers wait with anticipation to learn how Zarathustra was able to muster the energy and vitality needed to bite this monster and effect his transformation. We never hear it from him. It is only the animals that tell us what they know he knows—and in doing so make out of this insight what Zarathustra calls a "hurdy-gurdy song" (K4, 275; Z 332). It is all readers have. "You are the teacher of the eternal recurrence [*ewigen Wiederkunft*]—that is your destiny!" the animals tell him. This is what they say that Zarathustra must now know:

> The soul is as mortal as the body. But the knot of causes in which I am entangled recurs and will create me again. I myself belong to the causes of the eternal recurrence. I come again, with this sun, with this earth, with this eagle, with this serpent—*not* to a new life or a better life or a similar life: I come back eternally to the same, selfsame life, in what is greatest as in what is smallest, to teach again the eternal recurrence of all things, to speak again the word of the great noon of earth and man, to proclaim the overman again to men. (K4, 276; Z 333)

Although we do not know exactly how it comes about, we can see that the transformation represented in these words is one in which Zarathustra has learned to experience his own bodily being differently—as itself belonging to the *causes* of eternal recurrence. As a result, he need not fear that his message or his love will be extinguished; he need not be angry with his friends and followers for failing to hear. The same forces that have brought them into being have brought him into being as well. If they return eternally, so does he. His return, moreover, is what enables theirs. What Zarathustra has developed the ability to do, in short, is to experience the contradictions within his own love as occasions for him to learn more about himself, to deepen his knowledge of others, and to expand his love of humanity to include such moments of ambivalence. He has developed the perspective from which he can perceive the idea of eternal recurrence as an expression of the greatest love for life—a love so great that it encompasses all of life as necessary in its creative interconnectedness. The idea of eternal recurrence is an ideal, yet it is one that, according to Nietzsche, *dances*. It encourages love for bodily becoming and thus meets the criteria he taught his friends for remaining faithful to the earth.

Through this description of Zarathustra's transformation, Nietzsche enables readers as well to experience an apparent contradiction between the ideal of eternal recurrence and the ideal of the overman as a stimulus to their own creative becoming. On the one hand, the idea of eternal recurrence implies that everything that ever has been will be again; that everything that will be has already been. It represents a nauseating Dionysian wisdom that paralyzes action; it suggests the indifference of life to human beings. On the other, the ideal of the overman seems to celebrate an Apollinian freedom and power of individual self-creation. In Zarathustra's transformation, however, readers are able to experience both these ideals as expressions of a love for humanity that overcomes its own contradictions and remains faithful to the earth. For one who believes in eternal recurrence, it is absolutely necessary to take full responsibility for one's own bodily creativity in the making of every moment—what one creates is what will recur. For one who believes in the overman, on the other hand, it is absolutely necessary to discipline oneself to the currents of life flowing through one's body—that is where the sources of energy and wisdom are. As such, embracing eternal recurrence as an ideal expresses a love born of discipline to the rhythms of bodily becoming; embracing the overman expresses a love born of the capacity to see the necessity in all things. In short, having faith in either of these ideals demands having faith in the other. And such a double faith is a "daring dance," possible only for a body/self who is free from the need to escape suffering and strong enough to experience his experiences as occasions for ongoing knowledge and becoming. In every moment, we are taking what is given and creating the world in which we live.

Zarathustra, this time, does not respond to his animals' words. Instead, he closes his eyes and converses with his soul. The first words in his mind are: "O my soul, I taught you . . . to dance away over all Here and There and Yonder" (K4, 278; Z 333). Then his body/self beckons this soul who can dance to sing, and it sings "The Other Dancing Song."

In this song life again appears as a woman. Yet reflecting his magic transformation, Zarathustra's relationship to life has changed. His ability to embrace the ideal of eternal recurrence as an expression of his love for life brings him closer to dancing with (his image of) her.[15] As the song begins Zarathustra dances after her. He is "frantic to dance"; her rhythms impel his foot to move (K4, 282; Z 337). Yet she is elusive; and "rocking with dancing frenzy," hot in pursuit, Zarathustra, the rope dancer, falls. He is angry—angry with her, with his broken heart, and for falling. He takes out the whip he has not forgotten to bring saying: "You shall dance and cry" (K4, 284; Z 338).

Life responds. She defuses his anger by offering words that dance over the cleft between them: "Beyond good and evil we found our island and our green meadow—we two alone" (K4, 284; Z 338)—the green meadow, the dancing place, where perhaps life and Zarathustra will dance. Then life chides Zarathustra: "You do not love me nearly as much as you say; I know you are thinking of leaving me soon" (K4, 285; Z 338). The rift, life suggests, is not between life and Zarathustra, but within Zarathustra himself. He has not yet overcome his love-induced hatred of life. Yet in this "other" dancing song, Zarathustra *can* respond. He has redeemed his past, bit off the head of his great disgust, shattered the tombs of his youth, and embraced the doctrine of eternal recurrence. He has learned to love the overcoming that is (life in) himself.[16] He has learned, perhaps, how to dance.

Exactly what he whispers into her ear, again, we do not hear. Only the effect of his words is visible to us. He and life weep together: "then life was dearer to me than all my wisdom ever was" (K4, 285; Z 339). Their tears are tears of release—release of long-held fear, and a return to a sense of creative will, a discovery of a joy that is "deeper yet than agony" and that wants eternity (K4, 286; Z 339). In "The Yes and Amen Song" that follows, Zarathustra sings through his tears, proclaiming a love for life in a refrain that repeats seven times: "*For I love you, O eternity!*" (K4, 287–91; Z 340–3). In each of these seven "seals," he sings of a moment where joy desires its own eternity and thus affirms the knot of causes in which it is entangled. Dance appears as one of those joy-full moments. "If my virtue is a dancer's virtue . . . if this is my alpha and omega, that all that is heavy and grace should become light; all that is body, dancer; all that is spirit, bird—and verily, that is my alpha and omega: Oh, how should I not lust after eternity" (K4, 290; Z 342). To become a dancer, in this instance (again), is to undergo a magic transformation from which a person emerges with a love for life so strong that she greets with joy the thought that life recurs eternally.

In the end Zarathustra's change of heart remains a mystery for readers who read for the moral of the story. There are words and ideas but they are spoken by animals, whispered in the ear of life, or conveyed in songs. What philosopher can trust such sources? Yet Nietzsche's point is not that words cannot communicate, but that their meaning is not exhausted by their conceptual content. The value of eternal recurrence is not that it provides an ontological or cosmological account of creation, or a psychological test for one's actions. Eternal recurrence is an *ideal*—an ideal that expresses, for Zarathustra and Nietzsche, the all-encompassing love that the death of God makes possible. As such, its *meaning* lies in its ability to catalyze a change of heart in those who hear it—a change that requires that readers identify viscerally with Zarathustra, imitate his gestures, and discharge their nihilistic fears and fears of nihilism. It is a meaning that succeeds not by transmitting dancing words but by stirring in a reader a desire to dance—that is, a desire to love her bodily becoming, honor it, listen to it, discover it, discipline herself to it as the process of her own overcoming. Zarathustra describes such dance-enabled love as *holy selfishness.*

> And at that time it also happened—and verily it happened for the first time—that his word pronounced *selfishness* blessed, the wholesome, healthy selfishness that wells from a powerful soul—from a powerful soul to which belongs the high body, beautiful, triumphant, refreshing, around which everything becomes a mirror—the supple, persuasive body, the dancer whose parable [*Gleichniss*] and epitome [*Auszung*] is the self-enjoying soul [*selbst-lustige Seele*]. The self-enjoyment of such bodies and souls calls itself "virtue." (K4, 238; Z 302)

Learning the Lesson

In writing the Fourth Part of *Zarathustra*, Nietzsche follows the rhythms of love laid out in Parts One through Three. Going down, going up (First Part), going down (Second Part), going up (Third Part), Zarathustra must again go down to humanity to share this new understanding of love, the fruits of his ongoing overcoming. Waiting for this moment in his mountain cave, he hears a cry from the valley and goes down to investigate. In his wanderings he discovers the "higher men"—a group of individuals who have heard his teachings, taken them to heart, and are now, as were the rope dancer and himself, confronted by jesters of various shapes and sizes who threaten to overleap them. It is in teaching them that Zarathustra proves his ability to overleap lovingly his own youthful pride and loneliness.

The most significant mentions of dance in this part occur after Zarathustra has gathered the higher men into his cave. He invites them to a "Last Supper." Where Jesus urged his disciples to eat bread and drink wine

in memory of the body and blood that he shed for them, Zarathustra urges
his disciples to dance. At this meal he gives a speech, "On the Higher Man,"
that contains some of Nietzsche's most cited references to dance—including
the one Nietzsche himself quotes in his 1886 preface to *Birth of Tragedy*.

Zarathustra begins by sounding the theme of love and his own overcoming. He asks: "What has so far been the greatest sin here on earth?" He
responds: the sin of one who found no reasons on earth to laugh, for
"He did not love enough: else he would also have loved us who laugh" (K4,
365; Z 405). As if referring to his own experience in the Third Part,
Zarathustra continues: "he simply did not love enough: else he would not
have been so wroth that one did not love him. All great love does not *want*
love: it wants more. Avoid all such unconditional people! . . . they do not
know how to dance [*sie wissen nichte zu tanzen*]" (K4, 365; Z 405–6). The
point here is clear to those who have been following the path of Nietzsche's
dance imagery: to learn to dance is to learn to love oneself *enough* to love
life. It is to participate in the process of self-overcoming in a way that
kindles the ashes of one's faith in God into a love for humanity so great that
it transfigures one's experience of all present, past, and future loves. In his
own journey through the parts of this book, Zarathustra, the dancer, has
learned to dance.

That such a process of learning to love is described in terms of dancing,
moreover, illuminates a kind of extra-textual work that learning such love
requires of us, as readers. Seen as dancing, learning to love demands an attentiveness to and conscious participation in bodily becoming—the ongoing
creative process of metaphor-making and ideal-creation. It demands taking
responsibility for our physical health—for cultivating the energy required to
resist convention. It demands taking responsibility for how we educate our
senses—for cultivating an openness and awareness of our senses and holding
our ideals and gods accountable to their refinement of those senses. It
demands taking responsibility for being a nexus of desires, passions, and
instincts that jostle up against one another. It demands forgiving ourselves
and probing the physiological sources of our values. It demands being the
agent of our own redemption. As Zarathustra exhorts:

> You higher men, the worst about you is that you have not learned to dance as
> one must dance—dancing away over yourselves! What does it matter that
> you are failures? How much is still possible! So *learn* to laugh away over yourselves! Lift up your hearts you good dancers, high, higher! (K4, 367; Z 407)

Zarathustra is urging the higher men not to fall off the rope into nihilism or
despair even if their rivals appear to win. They must keep dancing by learning to laugh at themselves, knowing that their own failures are occasions for
developing the self-knowledge that enables love.

As the scene progresses, Zarathustra's actions confirm that he has learned how to do what he teaches—to laugh and to love. When the higher men invent a ritual and start worshiping an ass, Zarathustra is at first enraged. His anger booms forth as a love-induced contempt at the depths to which they are falling. Then one of the higher men confronts him—you taught us to laugh (K4, 392; Z 427). Zarathustra receives this wisdom as it is reflected at him. He laughs. The higher men were creating a festival. "I take that for a good sign: such things are invented only by convalescents" (K4 394; Z 429). Having thus demonstrated that he has overcome the fear-riddled love that attached him too tightly to the higher men, Zarathustra is free to descend his mountain and try, once again, to realize his love for humans.

Even so, and in spite of this book's length, it seems evident that *Zarathustra* is not finished. The narrative is both circular and open-ended. The conclusion of the Fourth Part returns to the point at which the Prologue began: Zarathustra is descending from his mountain with a heart overflowing with love for humanity. Further, it is clear that Zarathustra has still not fully realized his love and that he knows it. He is now looking for his *children*—not followers or friends, but those he can love who will love him. He has not yet found them. Nor, I would add, has he come to grips with the fact that in order to find his children he must learn to love a woman—not an ideal of woman, but one he can welcome as a co-creator in her own right.[17] That Nietzsche must have had some sense of this shortcoming is suggested by the refrain of the "Yes and Amen Song": "Never have I yet found the woman from whom I wanted children, unless it be this woman whom I love: for I love you, O eternity. *For I love you, O eternity*" (340). In Zarathustra's song, readers sense a rift that Zarathustra—and Nietzsche—has yet to overcome; "Eternity" cannot give him the children for which he yearns. Zarathustra needs to find a woman who can help him participate bodily in their creation. He is still in the process of learning to *dance*.

Forward and back

After *Zarathustra*, Nietzsche never wrote another fictional narrative. He had created his new sun, his dancing god. He had enacted his own transfiguring response to the death of God, expressed his love for humanity, and created an art he hoped would catalyze similar changes of heart in those who yearn to be rope dancers. In Nietzsche's opinion this book was his greatest, one he did not move beyond. What he writes in *Ecce Homo* of *Ecce Homo* is a sentiment that echoes throughout his later works: "I have not said one word here that I did not say five years ago through the mouth of Zarathustra" (K6, 373; EH 333).

Nevertheless, what Nietzsche writes in the years following *Zarathustra* looks and sounds different, and not just because Nietzsche returns to more philosophical styles. *Zarathustra* is Nietzsche's ideal—an image of what he himself desires to become, someone who believes in the overman and can embrace with joy the ideal of eternal recurrence. In the years following *Zarathustra*, then, Nietzsche writes to become more and more like Zarathustra, to learn by imitation how to make the movements of affirmation Zarathustra makes in the face of Nietzsche's own most abysmal thought: the eternal recurrence of Christian values. While some have claimed that after *Zarathustra* Nietzsche collapses into bitterness and vengefulness against a Christianity he is unable to escape, this reading suggests the reverse. What Nietzsche learns from writing *Zarathustra* is that he, like Zarathustra, must redeem his own past. Specifically, he must *redeem Christianity*—he must learn to love Christianity in the light provided by his own sun/son, Zarathustra/*Zarathustra*. As he writes in *Ecce Homo*: "The task for the years that followed [*Zarathustra*] was indicated as clearly as possible. After the Yes-saying part of my task had been solved, the turn had come for the No-saying, *No-doing* part: the revaluation of our values so far, the great war—conjuring up a day of decision" (K6, 350; EH 310).[18] The fact of *Zarathustra* for Nietzsche, his absolute joy in receiving it, funded a deep gratitude to the whole matrix of his life out of which it grew—a matrix that included the Christian faith against which he chafed. It is due to the experience of writing *Zarathustra* that Nietzsche hymns: "How could I fail to be grateful to my whole life?" (K6, 263; EH 221).

In the works of Nietzsche's last writing years, then, the No-saying and No-doing "in relation to Christian values" begins—but not as a simple negation. The task is *revaluation*: Nietzsche revisits moments of pain and suffering inflicted by Christian values in order to redeem them—that is, to affirm them as stimuli to the lives of free spirits everywhere.[19] Nietzsche loves Christianity by waging war against it—a war free of contempt. It is a war through which Nietzsche practices what Jesus teaches and learns to love his enemies. By loving Christianity as his enemy Nietzsche intends to demonstrate that in Zarathustra's love for humanity, Christian love overcomes itself. In these writings, Nietzsche calls upon "dance" once again, to figure a revalued act of incarnation. As in the vision of dance fore shadowed in *Birth*, God does not become human; human becomes a god who participates *bodily*, through practice and discipline, in the creation and realization of human love.

Chapter 3

Loving Life

The psychological problem in the type of Zarathustra is how he that says No and does No to an unheard-of degree, to everything to which one has so far said Yes, can nevertheless be the opposite of a No-saying spirit; how the spirit who bears the heaviest fate, a fatality of a task, can nevertheless be the lightest and most transcendent—Zarathustra is a dancer—how he that has the hardest, most terrible insight into reality, that has thought the "most abysmal idea," nevertheless does not consider it an objection to existence, not even to its eternal recurrence—but rather one reason more for being himself the eternal Yes to all things, "the tremendous, unbounded saying Yes and Amen" . . . But this is the concept of Dionysus once again.

—K6, 344–5; EH 306

Even though Nietzsche introduces Zarathustra's problem in this quotation as "psychological," the movement of the passage qualifies his thought. Learning how to affirm life in the face of one's "hardest, most terrible insights into reality" is a problem of learning how to dance. It involves engaging in practices that open a consciousness of bodily becoming—a physical consciousness—from whose perspective a person can greet the greatest challenges to her happiness with gratitude. It is the problem of how to effect the "magic transformation" that Nietzsche had observed years earlier in his study of Attic tragedy; and it is the problem that Nietzsche assumes for himself in relation to his Christian heritage in the books that follow *Zarathustra*, from *Beyond Good and Evil* (1886) and Book Five of the *Gay Science* (1887) to *Ecce Homo* (written in 1888) and the notes gathered and published posthumously as *Will to Power* (written in the 1880s). In these later works, Nietzsche writes to document how the suffering he discerns as perpetrated by Christian love has provided him with the necessary conditions for discovering the ideal of a life-affirming love that he creates Zarathustra, the dancer, to represent.

Amidst the wealth of these works, this chapter follows two strands that trace the fortunes of Nietzsche's dance imagery. In the first, Nietzsche pinpoints and details those aspects of Christianity whose revaluation Zarathustra's dance represents. Here I focus on the *Genealogy of Morals* where Nietzsche introduces the "ascetic ideal [*asketische Ideal*]" as the focus of his critique against Christian values. In this discussion, dance represents the acts of overcoming and revaluing the ideal of Christian love modeled on God's incarnation in Jesus Christ. *To dance is to love.*

Secondly, even though Nietzsche never advocates the practice of any particular kind of dance in these works, he lays out a host of criteria that any contender for the title of "dance" should meet. So carefully does Nietzsche attend to the care and well-being of bodies in these works, that they may be described as making a case for why the actual practice of dancing is integral to the project of revaluing values, and why it also serves as an illuminating image or ideal for the project as a whole. *To love is to dance.*

The implication of these two strands of discussion, taken together, is that, "dancers," in order to fulfill the role Nietzsche accords to dance in revaluing Christian love, must develop aesthetics of practice and performance that overcome the perception of dance encouraged by the ascetic ideal—a perception of dance as opposing religion. "Dance" must become what Nietzsche can imagine it to be: theopraxis, the bodily activity though which we generate and become our highest ideals. What appears in Nietzsche's later works, in short, is a call for dancers who can realize the potency of dance as art and religion— a call to which Duncan and Graham respond.

In following these strands, this chapter takes as a guide one of Nietzsche's strongest claims for dance, quoted on page 1 of this book from Book 5 of the *Gay Science*. Nietzsche's own ability to imagine dance as the *ideal*, the *art*, and the *only piety* of the philosopher, I argue, attests to his success in learning to love—and redeem—Christian morality.

Countering the Ascetic Ideal: *Genealogy of Morals*

Above all, a counterideal *was lacking*—until Zarathustra.

—K6, 353; EH 353

In the *Genealogy of Morals*, Nietzsche unleashes his sharpest and most incisive critique of Christian values. He also acknowledges a bottomless debt to Christianity and expresses his deepest gratitude. It was wrestling with Christian values that brought him to the place where he was open and ready to receive the gift of a counterideal [*Gegen-Ideal*] to the ideal of self-sacrificing love—Zarathustra, the dancer, the self-enjoying soul.

Written two years after completing *Zarathustra*, the *Genealogy* represents the most narrative of Nietzsche's later works. The long aphorisms in each of its three essays tell a story—a genealogy—of a particular term or nexus of terms.[1] His method in these essays is best described as physiological. He demonstrates how Christian values depend for their validity on physiological conditions and bodily practices that contradict what those values purport to represent. He writes: "every table of values . . . requires first a *physiological* investigation and interpretation, rather than a psychological one . . . for the problem of 'value *for what?*' cannot be examined too subtly" (K5, 289; GM I:17, 55). Specifically, in each essay, he approaches the problem of Christian love from a different angle. And in each case, his aim is to transform our experience of a value we take for granted as good and appreciate it as an expression of our physiological impoverishment. In doing so, however, it is not Nietzsche's intent that we fall off the rope into nihilism and despair. He wants to teach us to make the right steps—the movements of love in relation to our sickness. He writes to transform our experience of our own sickness, so that we can embrace it as a morning sickness, and an indication of our potential to give birth to new ideals of highest value—dancing stars—and to ourselves as creators of value. Even though there are few mentions of dance in this book, their placement illuminates why Nietzsche chose dancing as an effective symbol for this revaluation of the highest Christian value, love.

Revaluing Revaluation

In the first essay of the *Genealogy*, Nietzsche introduces a comparison between modes of valuation similar to that launched years earlier in *Birth*. Again he charts a relationship between the two modes of valuation and characterizes that relationship as representing a reversal. Here he names the movement from one mode of valuation to the other a "revaluation of values." Associating this revaluation with the Jews (rather than Socrates) Nietzsche describes Christian love as its fullest bloom: in the ideal of self-sacrificing love, the "priests" secured a means for acquiring power over the stronger "nobles." The priests did so by creating a value of their weakness. As Nietzsche writes: "This Jesus of Nazareth, the incarnate gospel of love, this 'Redeemer' who brought blessedness and victory to the poor, the sick, and the sinners—was he not this seduction in its most uncanny and irresistible form, a seduction and bypath to precisely those *Jewish* values and new ideals?" (K5, 268–9; GM I:8, 34). As Nietzsche interprets Jesus *is* the "incarnate gospel of love"; Jesus incarnates God's love in the form of self-sacrifice. For those who believe in Jesus, God demonstrated how much he loves humans by becoming human (in Jesus) and then sacrificing himself in order to redeem humans from their sin. So too those who follow Jesus, Nietzsche discerns, are encouraged to cultivate a similarly unselfish love, sacrificing their bodies and selves for God—and for reward in heaven. Nietzsche's picture of "Christianity" is admittedly a sketch. Yet it is one steeped in a deep understanding of scripture and tradition. Like a good caricaturist, he captures in a few broad strokes defining features of a faith that many people who call themselves Christians share.

Several points in Nietzsche's discussion of revaluation bear upon the significance of dancing in his project. First, revaluation is not new. Nietzsche does not claim to invent the term; he discovers it. What he discovers, through an etymological study of the words used to designate "good" and "bad" and "evil," is that historical shifts have occurred in which what was "good" has become "evil" and what was "bad" "good." The meanings of these values, moreover, do not evolve in a logic-driven or linear fashion; they arise to serve the physiological needs of those who hold the values. In naming his own project as one of revaluation, then, Nietzsche is acknowledging his debt to the very tradition he critiques: he has isolated the secret of its power and turned it against itself. He will revalue revaluation to serve his own physiological needs.

Second, Nietzsche roots the impulse for revaluation in an inability to accomplish what he has used "dance" to represent in earlier works. As he admits, revaluation is a creative act; it is an expression of the will to power. The problem with the first revaluation is not the revaluation per se; but, as Zarathustra taught, that the will that creates such values expresses its own impoverishment. Of the priests Nietzsche confirms: "It is because of their impotence that in them hatred grows to monstrous and uncanny proportions, to the most spiritual and poisonous kind of hatred" (K5, 266–7; GM I:7, 33). This impotence-impelled hatred is *ressentiment*. The priests are unable to *dance*: they cannot discharge their emotions, honor their kinetic responses, or participate in their bodily becoming. They seek revenge in the only way open to them: the mental act of changing the meaning of words, manipulating linguistic metaphors at a secondary level of metaphor-making. Nietzsche summarizes:

> The slave revolt in morality begins when *ressentiment* itself becomes creative and gives birth to values: the *ressentiment* of natures that are denied the true reaction, that of deeds, and compensate themselves with an imaginary revenge. While every noble morality develops from a triumphant affirmation

of itself, slave morality from the outset says No to what is "outside," what is "different," what is "not itself"; and *this* No is its creative deed. This inversion of the value-positing eye—this *need* to direct one's view outward instead of back to oneself—is of the essence of *ressentiment*: in order to exist, slave morality always first needs a hostile external world; it needs, physiologically speaking, external stimuli in order to act at all—its action is fundamentally reaction. (K5, 270–1; GM I:10, 36–7)

In Nietzsche's view, *ressentiment* drives the ideal of self-sacrificing Christian love. It is what the ideal of Christian love expresses. It is because the priests are physiologically frustrated that they create an ideal of love that vilifies their enemies as evil and sanctifies their own powerlessness as holy—a love that seduces the strong to will their own sickness and suffering as the path to communion with God. The "incarnate gospel of love" encourages and demands hatred of the body because those who created these values lacked the strength to honor their own bodies.

That *ressentiment* lies at the root of Christian love, then, suggests what revaluation must entail. Christian love expresses a contradiction: it always already expresses hatred. As such, revaluation can never be a mere reversal in which what was loved is now hated and what was hated is now loved. A new ideal of love is required, one that extends to include all of life, Christian hate included, as necessary. Such an ideal, for Nietzsche, is the idea of eternal recurrence.

Third, in Nietzsche's description of the two modes of valuation, it is not surprising where the image of dance appears: as an activity of those he describes as "nobles." The nobles danced; and they did so in order to engender the health and strength needed to sustain their life-affirming values. As Nietzsche explains: "The knightly-aristocratic value judgments presupposed a powerful physicality, a flourishing, abundant, even overflowing health, together with that which serves to preserve it: war, adventure, hunting, dancing [*Tanz*], war games, and in general all that involves vigorous, free, joyful activity" (K5, 266; GM I:7, 33). This placement of dance might seem odd at first. Its value in this passage does not lie in the fact that it is an art, but that it is a kind of art that engages the creative powers that animate all dimensions of our bodily lives. Dance is "vigorous, free, joyful activity." It serves to *preserve* the *powerful physicality* that the nobles' life-affirming values presuppose. In other words, for the nobles, dancing provides an effective means of preserving their capacity to create their own values. Dancing *expresses* the energy human beings need in order to discern when to fight against whatever threatens to impoverish existence or bind their spirits. By contrast, the priestly ideal of self-sacrificing love expresses a body that yearns to create beyond itself but cannot—a body spirit that cannot *dance*.

This placement of dance in the First Essay suggests its significance in Nietzsche's overall project of revaluing Christian values. In so far as the first revaluation of values preached passivity and weakness over and against this value-creating strength, the sovereignty of the Christian ideal of love has depended upon abolishing dance from religion wherever it has occurred. Dancing, as a vigorous, free, joyful activity, would provide individuals with the opportunity to sense and discover the "power" of their own physicality and thus reject the values of self-sacrifice and denial. Furthermore, this placement also suggests that the primary force driving antipathy to dance in Christian contexts is not hatred of the body per se but the ideal of self-sacrificing love itself. Antipathy to dance and to bodies ensures that persons will remain dependent upon and faithful to the otherworldly salvation promised by the Christian narrative. In short, to the extent that they believe that dancing must be suppressed, Christian authorities betray what they deny: their values do not express eternal truth or reality but rather stand or fall based on the training to embodiment that persons in community receive.

Furthermore, this placement of dance in the *Genealogy* supports the reading of *Zarathustra* offered in chapter 2: the creation of new values in the wake of God's death cannot occur until and unless people cultivate a different relationship to their bodily being than that perpetuated under the mantle of self-sacrificing love. Values of unselfish love born of *ressentiment* have densed our senses, numbed our receptivity to experience, and predisposed us to certain conclusions. We have become the incarnate gospel of love. Our bodies have grown weaker, less capable of creating beyond themselves, while our minds have grown richer and more complex.

Nevertheless, in the *Genealogy* as noted, Nietzsche's intent is to redeem Christianity—to will backward and learn to love it. He does so as he appreciates the value of these richer and more complex minds. With them comes as well an ability to reflect back on the first revaluation and contest its value. It is in such minds, Nietzsche avers, that the desire for life-affirming values awakens. Such free spirits become a "battleground" for alternative modes of valuation: they resist the life-negating values that enable them to think critically about values in the first place. As Nietzsche affirms: "[T]oday there is perhaps no more decisive mark of a '*higher nature*,' a more spiritual nature, than that of being divided in this sense and a genuine battleground of these opposed values" (K5, 286; GM I:16, 52).

Here the project of revaluing Christian values acquires a new twist: revaluation must involve overcoming this contradiction—this Zarathustrian, Dionysian rift or cleft—within ourselves. And overcoming this contradiction requires not only reinventing practices that preserve a powerful physicality, but opening ourselves once more to the primary levels of our own bodily creativity. We must learn to embrace and love this contradiction as the necessary condition for creating values that enhance rather than deplete human well-being. Dance, as an activity that is both physically vigorous and symbolic, both bodily and reflexive, stands poised to emerge as an aid to free spirits engaged in such a project. Again Nietzsche will call upon dancing to figure the transformation in our experience of bodily being that provides us with a physiological-spiritual perspective from which to love life.

An Inward Turn

The Second Essay, treated only briefly here, amplifies this argument for why dancing emerges in Nietzsche's work as the figure for a revalued Christian love. In the Second Essay, Nietzsche examines the sickness of "bad conscience": the will to inflict violence on oneself, especially in the forms of guilt and in expectations of unselfishness. Nietzsche's analysis parallels that of the First. The will to power of people living in society, their "instinct for freedom," cannot discharge itself. Where the priests responded by creating an enemy Other, here people respond by turning their hatred inward. "This *instinct for freedom* [*Instinkt der Freiheit*] forcibly made latent . . . pushed back and repressed, incarcerated within and finally able to discharge and vent itself only on itself: that, and that alone, is what the *bad conscience* [*schlechte Gewissen*] is in its beginnings" (K5, 325; GM II:17, 87). This turning inward of frustration against oneself finds expression in values of selflessness: "only the bad conscience, only the will to self-maltreatment provided the conditions for the *value* of the unegoistic" (K5, 326–7; GM II:18, 88). By this route Nietzsche again arrives at the ideal of Christian love: the "paradoxical and horrifying expedient that afforded temporary relief for tormented humanity, that stroke of genius on the part of Christianity" was to presume that God sacrifices himself for the guilt of mankind out of *love* (K5, 331; GM II:21, 92). Christian love, in other words, is not just predicated on hatred of others or of one's weak body, but on hatred of one's very *self*. Such a teaching, for Nietzsche, represents an extreme act of "conscience-vivisection and self-torture [*Selbst-Thierquälerei*, or "cruelty to animals"]," arts which we "modern men" are experts at practicing (K5, 335; GM II 24: 95).

At the close of this essay, Nietzsche makes the allusion that connects the discussion with his images of dance. He appeals to *Zarathustra* as offering an ideal of great love and contempt that can redeem us from this cycle of self-destruction. The way in which Christian love fails to nourish a good and healthy conscience attests to what we need to learn from Zarathustra's dance: how to *love ourselves* enough to hate the ways in which we inflict violence upon ourselves and limit our creative bodily potential. We must do so through discipline—by developing the physical consciousness and health we need to reject the guilt we are practiced at administering to ourselves and others. We need to learn a holy selfishness, a self-discipline that is an expression of self-love. Like Zarathustra, we need to learn to dance.

Note that Nietzsche is not suggesting that it is possible or even desirable to undo the effects of this inward turn he critiques. He does not claim that inflicting violence upon oneself is inherently bad. On the contrary, such reflexive movement is the work of life or "spirit" creating beyond itself. Consciousness and language, beauty and morality, are some of its fruits. Instead, Nietzsche calls us in this essay and the next to discern nuances in the kind of violence at work, in the rationale given for its application, and in the ends that it serves. In so far self-inflicted violence serves to maintain belief in the reality of an other or ideal world, it renders us increasingly incapable of responding creatively and affirmatively to challenges in our lives. To take steps to the overman, by contrast, means that a person must learn to distinguish between ways of relating to herself that foster an ongoing becoming (and the violence such becoming involves) and those that immure her in idea-edifices. She must learn to distinguish self-discipline from self-denial, and self-love from self-indulgence.

Countering Ascetic Ideals

In the Third Essay, Nietzsche's extended critique of the "ascetic ideal" fleshes out techniques of violence on which Christian values depend, and to which Zarathustra's dance offers a counterideal, a theopraxis. This essay poses as an extended exegesis on a sentence from "On Reading and Writing"—the speech in which Zarathustra claims that he could believe only in a god who could dance and that now a god dances through him. Here, as Nietzsche writes, Zarathustra dances through him: Nietzsche demonstrates the kind of reading of his writing he desires.

In this essay Nietzsche invents a category for the kinds of ideals he has been describing since *Birth*—ideals that express and encourage physiological impoverishment. He coins the term "ascetic ideal" to refer to values that do not *dance*—that presume as natural and/or sanctify as salvific a split between the body on the one hand, and the mind, spirit, or soul, on the other. As *ideal*, an ascetic ideal is a product of human creation, expressing a multi-level process of metaphor creation, and offering a picture of ourselves as who we are and should be. As *ascetic*, however, these ideals deny their own origin in bodily movement. Predicating radical distinctions between mind and body, doer and deed, or God and human, ascetic ideals function to reinforce a sense of self as a thinking "I" operating over and against the senses, passions, and instincts of the impotent body in which "I" dwell. They isolate that "I" as the locus of agency and causality. By the logic of the ascetic ideal, then, I am responsible for my suffering. In this way the ascetic ideal can be used to justify acts in which I inflict suffering on myself (as in asceticism) or on others (as in war) in amputating, extinguishing, or otherwise denying bodily sensibility. In all, while an ascetic ideal may mean something different to the artist (Nietzsche cites Wagner), to the philosopher (Kant and Schopenhauer come to mind), and to the priest (Christian) and his devotees, three constants prevail: an ascetic ideal justifies a physiological impoverishment of life; it acquires power as an object of belief through technologies and practices that teach people to ignore their bodily becoming as having anything to offer the pursuit of truth, knowledge, consolation, or even health; and it embodies a self-contradiction. In promoting the denial of life, an ascetic ideal weds a person ever more tightly to his existence as having meaning.

The example most relevant for a discussion of Nietzsche's dance imagery is that of the priest and his devotees. The priest, ostensibly Christian, purports to respond to the suffering of those who are weak—"the deep depression, the leaden exhaustion, the black melancholy of the physiologically inhibited" (K5, 377; GM III:17, 130). The priest responds, however, not by tackling the cause of the malaise; he rather works "the self-destruction of the incurable" and directs "the *ressentiment* of the less severely afflicted sternly back upon themselves" (K5, 375; GM III:16, 128). To this end the ascetic priest invents techniques, both innocent and guilty, that entwine the themes of revenge and guilt Nietzsche has introduced in the first two essays and does so in ways that educate followers to desense or oversensitize their bodily sensation. His act, while creative, is one in which life turns against itself with an unhealthy violence. The priest makes his followers sicker and more dependent upon him and his techniques for their relief.

The difference between the innocent and guilty techniques Nietzsche describes traces degrees of harm the techniques inflict on the ability of a body to create beyond itself. The means Nietzsche deems "innocent" involve the "hypnotic muting of all sensitivity, of the capacity to feel pain—which presupposes rare energy and above all courage, contempt for opinion" (K5, 382; GM III:18; 134); as well as the mechanical activity of work, especially in helping others and nurturing the health of the herd (K5, 382–4; GM III:18, 134–6). The "guilty" techniques, in contrast, work to shatter the nervous system, by submitting persons to painful "orgies of feeling" (K5, 388; GM III:20, 140). As Nietzsche explains, the aim is "To wrench the human soul from its moorings, to immerse it in terrors, ice, flames, and raptures to such an extent that it is liberated from all petty displeasure, gloom, and depression as by a flash of lightning" (K5, 388; GM III:20, 139). In either case, the prescribed actions enforce faith in an otherworldly soul over and against the bodily world as the locus of knowledge, truth, and consolation. As such, these actions sustain the kind of ideals Zarathustra rejects as hostile to life.

Yet it is from studying these practices, Nietzsche admits, that he comes to understand what he needs to learn in order to revalue Christian values and resist their life-negating effects. What he learns from his study of the ascetic ideal is that Christianity does not primarily concern what people believe. Christian faith reproduces itself through practices. Of the many implications of this insight, one in particular stands out: Christian values cannot be refuted with logical argument. Contesting the effects of Christian values requires finding trajectories within our own bodies and within history itself where Christianity, as one more part of life, *is overcoming itself.*

A first glimpse of such a trajectory appears in Nietzsche's account of a "lucky hit" in Christian history: a rare case where ascetic ideals catalyzed a powerful coordination of body and soul among those who refused to abide by them (K5, 367; III:14, 121). Nietzsche explains:

> Man has often had enough; there are actual epidemics of having had enough (as around 1348, at the time of the dance of death); but even this nausea, this weariness, this disgust with himself—all this bursts from him with such violence that it at once becomes a new fetter. The No he says to life brings to light, as if by magic, an abundance of tender Yeses; even when he *wounds* himself, this master of destruction, of self-destruction—the very wound itself afterward compels him *to live.* (K5, 367; GM III:13, 121)

Nietzsche appeals here and later in this essay (K5, 391; GM III:21, 142–3) to the dance epidemics in Christianity that he introduced in *Birth* as examples of a Dionysian impulse at work. In these epidemics the dancing embodies a contradiction in such a way that it impels a transformation in the experience of those who dance. The dance gives symbolic bodily expression to suffering and sickness in such a way that the people dancing find in their suffering a reason to love life. As Nietzsche explains, the dancing erupts "in the wake of repentance and redemption *training*"; it presupposes training in self-denial. At the same time, as a result of that training, people experience a disgust with themselves so great that it explodes, animating their entire body-making symbolic and affording them a new experience of themselves as creative. The dancing is thus *therapeutic.* It discharges the feelings of disgust and provides a new physiological perspective from which to feel love for themselves, for life, and even for their sickness.

Several points are worth noting. First, in this example Nietzsche places dance in Christian history. "Dance" is not opposed to Christianity in a dualistic sense; their histories are entwined. Second, while most Christian authorities, as he admits, have succeeded in arresting the dance and, harnessing people's senses to ideals of self-sacrificing love, the authorities very success foreshadows their fall. Their success presupposes what a dancer knows: that it is possible to educate the senses; that physiological health finds expression in ideals; that ideals become bodies, and that this whole cycle occurs in and through a body's own movement. Thus, even though Nietzsche rails against the ascetic ideal and accompanying techniques, he also acknowledges in the ascetic ideal the seeds of Christianity's own over-coming. The ascetic priest exploits (while denying) a logic of bodily becoming that Nietzsche is able to honor and reclaim as an enabling condition for creating *new* ideals. In other words, the act of detailing the horrifying con-tradictions within the ascetic priest's position gives Nietzsche great hope: it provides him with resources for imagining an alternative mode of valuation that acknowledges and honors not a particular ideal of good but the process of bodily becoming as the locus for creating life-affirming ideals—gods that could dance. In the ascetic ideal, then, Nietzsche not only isolates the offending element of Christian values, he identifies it as that moment where Christian values are overcoming themselves *in him*.

In the remainder of the Third Essay, Nietzsche defends his thesis by evaluating other aspects of modern culture that appear to counter the work-ings of the ascetic ideal. Specifically, he addresses the claim that science provides a rational ground from which to expose hypocrisies of religion. Even his own free-spirit science, Nietzsche admits, expresses the physiolog-ical impoverishment—"the affects grown cool, the tempo of life slowed down . . . seriousness imprinted on faces and gestures" (K5, 403; GM III:25, 154)—characteristic of the ascetic ideal. Science worships the ques-tion mark as God and assumes that since what it knows fails to satisfy its desire for knowledge, the problem is knowledge (K5, 405; GM III: 25, 156). Rather than creating new values, science reinforces faith in the ascetic ideal: in seeking to debunk Christianity as offering false claims to truth, sci-ence implies that truth is what counts. As Nietzsche confirms: we "still derive *our* flame from the fire ignited by a faith millennia old, the Christian faith, which was also Plato's, that God is truth, that truth is *divine*" (K5, 401; GM III: 24, 152). Of so-called free spirits he summarizes: "That which *constrains* these men . . . this unconditional will to truth, is *faith in the ascetic ideal itself*, even if as an unconscious imperative—don't be deceived about that—it is the faith in a *metaphysical* value, the absolute value of *truth*, sanctioned and guaranteed by this ideal alone" (K5, 400; GM III: 24, 151).

The predicament Nietzsche elaborates here is his personal version of Zarathustra's challenge. The greatest threat to Nietzsche's hopes for creating life-affirming ideals—his own most abysmal thought—is that the science that has enabled him to develop a critical purchase on the ascetic ideal at work in Christian love itself perpetuates faith in that ideal. To learn to dance like Zarathustra, then, Nietzsche must learn how to experience this contradiction (i.e., being both the ascetic ideal and its opposite) as a reason to love life. He does so by willing backward: he acknowledges how his experience of ascetic ideals has propelled him to study their bloody history, unmask their physiological causes, call into question their value, and create his own dancing sun. As he avers: "Which of us would be a free spirit if the church did not exist?" (K5, 270; GM I: 9, 36).

One implication of this analysis, however, has troubled readers for years. What does it mean to affirm torture of various kinds as necessary? If the ascetic ideal is ubiquitous in culture, are we doomed to repeat eternally the patterns of self-denial with which we are familiar? Are we forever trapped within our thinking I's, destined to neglect our sensory education, and perpetuate the crimes for which we charge Christian morality?

Nietzsche's response is no. What we need, however, is not a better argument for a better truth. What we need is what he claims to have provided in *Zarathustra*: "*Art* . . . in which precisely the *lie* is sanctified and the *will to deception* has a good conscience, is much more fundamentally opposed to the ascetic ideal than is science" (K5, 402; GM III: 25, 153–4). *Zarathustra* as the "good will to appearances" counters the workings of the ascetic ideal by transforming our experiences of Christian morality and God's death from horror to deep gratitude and joy. *Zarathustra* gives us Zarathustra—a dancing, singing image of ourselves as creators of value. It stimulates us to find within ourselves the capacity to imitate Zarathustra's gestures and learn to love humanity as he does. It is thus designed to awaken a physical consciousness of ourselves as bodies generating and becoming their highest ideals. By reawakening this dimension of human creativity, Nietzsche aims to catalyze a new physiological health, enabling people to embrace the contradiction between alternate values as a stimulus and guide in creating new ones. The different points of value define a trajectory forward.

In summary, this analysis of the *Genealogy* illuminates the significance for Nietzsche's revaluing values of Zarathustra's dancing. That Zarathustra *dances* signals that his self-enjoying love for humanity *is* the ideal in which Christian self-sacrificing love overcomes itself. Although we never see Zarathustra dance (except as he describes in his song), we are led to believe that the fact that he knows how to dance—the secret of his dance-like walk—is what enables him to endure seven days of darkness and emerge with the ability to will backwards. He resurrects the tombs of his youth, and learns to love himself enough. Moreover, the fact that we cannot watch his health-enabling dance is itself significant. By not revealing the forms of Zarathustra's movements, Nietzsche expresses his own love for humanity. He refuses to offer readers a formula to which they must conform; he loves us too much. Instead, we must seek solitude, learn to wait, cultivate our own physical consciousness, and give birth to ourselves as creators. We must rely on resources—physical, spiritual, intellectual—that our own practices of dancing have stirred in us.

As the dancer described in Book 5 of *Science*, then, Zarathustra practices a form of *art* creation that is also a self-creation; he enacts an *ideal* of ideal making that remains accountable to the health of the bodies through which that ideal making occurs; he expresses a *piety*, a service of God, by honoring and overcoming in himself the effects of the ascetic ideal. He realizes a Dionysian faith and love. As Nietzsche confirms, "My concepts of the 'Dionysian' here became a *supreme deed*" (K6, 343; EH 304). Zarathustra incarnates a theopraxis best represented as "dance." In Zarathustra, Nietzsche thus incarnates a vision of dance as theopraxis.

A Physical–Spiritual Discipline: The Works of 1888

> Not to make men 'better,' not to preach morality to them in any form, as if 'morality in itself,' or any ideal kind of man, were given; but to create conditions that require stronger men who for their part need, and consequently will have, a morality (more clearly: a physical-spiritual discipline) that makes them strong!
>
> —WP #981, 513

In his works of 1888—*Case of Wagner*, *Twilight of Idols*, *The Antichrist*, *Ecce Homo*, and *Will to Power*[2]—Nietzsche fleshes out the relationship to bodily becoming, the "physical-spiritual discipline," that he uses Zarathustra's dancing to represent. Skipping among Nietzsche's scattered references to dance, movement, and bodily becoming in these later texts discloses the shape of such a discipline—a vision for what dance can and should be for free spirits. *A vision for dance as theopraxis.* In turn, the vision of dance that emerges reflects back on the individual mentions of dance, providing additional support for the idea that Zarathustra's dance not only expresses a revalued Christian love, but provides a guide for the process of revaluation in general as well.

A Physiological Contradiction

In the works of 1888, Nietzsche grapples with the predicament that his study of the ascetic ideal helped him to articulate at the end of the *Genealogy*: "modern man" is the site where competing modes of valuation war against one another. It is a predicament that Nietzsche describes as a "physiological contradiction [*des physiologischen Widerspruchs*]" in *Case of Wagner* (K6, 48; CW 187) and as a "physiological self-contradiction [*der physiologischen Selbst-Widerspruch*]" in *Twilight of Idols* (K6, 143; TI 545).[3] We "modern men" who pursue truth via rational means live this physiological contradiction because we both continue Christian morality (in maintaining faith in truth) and oppose it (in our rational pursuits). This contradiction, moreover, is not an intellectual problem: it lives in our bodies. It has become us. As Nietzsche writes: "all of us have, unconsciously, involuntarily in our bodies values, words, formulas, moralities of *opposite* descent—we are, physiologically considered, *false*" (K6, 53; CW 192); "Biologically, modern man represents a *contradiction of values*; he sits between two chairs, he says Yes and No in the same breath" (K6, 52; CW 192). This understanding of himself is what his experience of receiving and writing *Zarathustra* has allowed Nietzsche to comprehend: he bears an ongoing debt to Christianity in his very opposition to Christian values.[4]

Nevertheless, as suggested at the end of the *Genealogy*, the problem in Nietzsche's view is not that we are this physiological contradiction, but that we do not have the strength to sustain it. We refuse to perceive the opposition in ourselves as an opposition. In so far as we feel the tension of this antithesis, we seek means for relieving it. We choose one faith to the exclusion of the other; we gloss over the difference or map the difference onto discrete realms of life. Yet as Nietzsche confirms: "These opposite forms in the optics of value are *both* necessary: they are ways of seeing, immune to reason and refutations. One cannot refute Christianity; one cannot refute a disease of the eye"; rather, "[w]hat alone should be resisted is that falseness, that deceitfulness of instinct which *refuses* to experience these opposites as opposites" (K6, 51; CW 191).

This framing of the problem carries implications for what revaluation entails. Again, revaluing Christian values cannot involve making a mental choice to abide by one set of values to the exclusion of the other. What must be revalued is precisely this perception of morality as a function of mental choice. What must be revalued is the split between body and mind presumed by such an understanding of revaluation. What is wrong is not religion or science per se, but a religion or science that denies the conflict between them as generative of each. What is wrong is the *ascetic ideal* at work in our perception and use of art, religion, and science. What is wrong is the attempt to reduce life to one dimension, for "Above all, one should not wish to divest existence of its *rich ambiguity*" (K3, 625; GS #373, 335).

Given the nature of what must be overcome, then, the task of revaluation involves transforming our experience of our own physiological contradiction, our own Yes and No. We need "The strength to withstand tension. The width of tensions between extremes" (K6, 138; TI 540). We must learn how to pursue science as an art, and appreciate art, religion, and metaphysics as participating in a play of appearances that enables science. Only then will we live the tension between them without falling into either nihilism or despair. By learning to move between these contrasting perspectives on life, in our own "daring dance," we can give rise to new values in art, science, and religion that honor and nourish our ongoing bodily becoming. We will learn to experience this physiological contradiction as the condition for our freedom, our knowledge, and our ability to love.

In Nietzsche's later works, dancing appears as activity capable of serving this function in at least four respects that are by now familiar. *Dancing* is (1) a discipline for self-knowledge in which a person (2) strengthens the instincts; (3) educates the senses, and (4) invigorates the energy needed to embrace metaphor-making as a creative, bodily process. In doing so, dancing emerges as a criterion for the kind of activities to which our acts of reading and writing must be held accountable.

A Discipline of Self-knowledge

Through the long succession of millennia, man has not known himself physiologically: he does not know himself even today.

—WP #229, 132

The dancing Nietzsche imagines as a discipline for self-knowledge is one that teaches people through action about dimensions of their bodily selves that fidelity to the ascetic ideal teaches them to ignore. Several references to dance in *Beyond Good and Evil*, written in 1886, acknowledge this disciplinary aspect. Here Nietzsche represents dance as an activity that is not only symbolic, but requires training, years and even generations of practice. It is a kind of training for which there is no substitute. There is no way to reap the rewards any other way than by doing the work oneself. The training that dance represents is a process that effects a bodily transformation through which each person must pass on his own, through his own initiative, by calling his consciousness to attend to the patterns of the required physical coordination. As Nietzsche writes: "[T]he curious fact is that all there is or has been on earth of freedom, subtlety, boldness, dance [*Tanz*], and masterly sureness, whether in thought itself or in government, or in rhetoric and persuasion, in the arts just as in ethics, has developed only owing to the 'tyranny of such capricious laws'; and in all seriousness, the probability is by no means small that precisely this is 'nature' and 'natural'—and *not* that *laisser aller*" (K5, 108; BGE V:188, 100).

Nietzsche's points here illuminate the kind of self-knowledge he imagines dancing to provide. First, Nietzsche debunks the idea that what is "natural" is an unfettered release of bodily energies. What is "natural" is precisely a process of self-reflexive becoming that follows laws. And it is only in this second sense that dance is "natural." At the same time, neither are these "laws" themselves natural or rational; they are "capricious" in the sense that they emerge out of the training process itself as its conditions. Dancing, then, helps us discover laws for ourselves that are not grounded in a foundation other than the ongoing rhythms of our sensory, experiential bodily becoming—laws that enable our ongoing bodily becoming.

Second, Nietzsche acknowledges that the process of discovering and following these laws is itself productive: something is created—a masterly sureness, a freedom and boldness. The act of practicing transforms that agent who practices into a dancer. While its placement in the above list might suggest that dancing is only a metaphor for what occurs in the areas of culture he proceeds to mention, Nietzsche does include the arts as one area. As such, dance appears twice. This double placement implies that all cultural forms share to some extent in the logic of bodily becoming Nietzsche's uses dance to represent. In so far as they do, they succeed in producing qualities and ideals that Nietzsche values. They serve human living. Thus Nietzsche does not reduce dance to a metaphor for the movement of thoughts or pens, but rather frames the movement of thoughts and pens as stylized instances of bodily practice. This passage, in short, debunks those who interpret Nietzsche's dance imagery as representing a call for *laissez allez*—a free abandonment to the whims of sense and instinct. For Nietzsche, in fact, such interpretations of dance or of freedom represent evidence that human instincts for self-preservation have degenerated to the degree that humans are no longer capable of discerning what is good for them (K6, 143; TI 545–6).

In a second example from *Beyond Good and Evil*, Nietzsche elaborates on how dancing enables self-knowledge. He confirms: "What is essential 'in heaven and on earth' seems to be, to say it once more, that there should be *obedience* over a long period of time and in a *single* direction: given that, something always develops, and has developed, for whose sake it is worth while to live on earth; for example, virtue, art, music, dance [*Tanz*], reason, spirituality—something transfiguring, subtle, mad, and divine" (K5, 108–9; BGE V:188, 101). In this list of things that make life worth living, dancing appears alongside reason and virtue as representing obedience and investment in one direction over time. Due to this obedience, something "develops"; and in the case of dance, what develops is what impels Nietzsche to identify dancing in the passage above as a paradigm for these other forms. As he noted in *Human*, the particular obedience that the practice of dancing requires provides us with a unique knowledge of ourselves as processes of bodily becoming, as capacities for becoming the metaphors we create. We thus know, in a way we otherwise could not, the physiological conditions of our own freedom: we know that the source of our freedom lies in the self-overcoming movement of bodily being. Nietzsche is not saying that dancing allows us to know the "real" body as a material thing or object. He is saying that "the body" is such a "thing" that we cannot know it unless we participate in the rhythm of self-becoming that it is, and do so with attentive obedience. The knowledge of ourselves that dancing affords thus serves us as a criterion for evaluating our courses of action.

This sense of dance as a discipline of self-knowledge hovers silently in moments when Nietzsche suggests that the kind of thinking he endorses is "something light, divine, closely related to dancing and high spirits" (K5, 147; BGE VI:213, 139). As noted in a passage from *Twilight* cited earlier, Nietzsche recommends that thinking be taught as a kind of dancing. He continues: "For one cannot subtract dancing in every form from a noble education—to be able to dance with one's feet, with concepts, with words: need I still add that one must be able to do it with the pen too—that one must learn to *write*?" (K6, 110; TI 512–3). While commentators notice that concepts, words, and pen dance, Nietzsche begins with the phrase "with one's feet." In doing so, he is reminding us what it is about dancing that makes it such a powerful metaphor for writing and thinking in the first place. Rather than imply that the act of dancing with one's feet is not important, this passage implies that the act of dancing with one's feet is necessary in order to develop the kind of self-knowledge needed to think and write with subtlety, boldness, and a sense of mastery. Dancing is powerful as a metaphor for thinking because it represents a bodily practice of attentiveness to and conscious participation in bodily becoming; because it demands obedience applied in a particular direction over time. Nietzsche strings these references together so that dancing, as a metaphor, does not lose its power to affect the senses. He is serious: dance cannot be subtracted "from a noble education" and not just because it preserves a powerful physicality. Dancing is a discipline for consciousness—a discipline in which we generate *physical consciousness*: knowledge of ourselves as participating in our own rhythms of bodily becoming.

The kind of self-knowledge that the discipline of dancing generates can best be described by harking back to Nietzsche's description in Book 5 of *Science* of "the great health."

The great health.— . . . Whoever has a soul that craves to have experienced the whole range of values and desiderata to date . . . whoever wants to know from the adventures of his own most authentic experience how a discoverer and conqueror of the ideal feels, and also an artist, a saint . . . needs one thing above everything else: the *great health*—that one does not merely have but also acquires continually, and must acquire because one gives it up again and again and must give it up. (K3, 635–6; GS #382, 346)

Such health, "a new health, stronger, more seasoned, tougher, more audacious, and gayer than any previous health," is what free spirits enjoy; it is what overflows in their "daring dance." It represents the "pleasure and power of self-determination," the vitality and energy needed to "dance" over habits and conventions, "dancing even near abysses" (K3, 583; GS #347, 289–90). The implication of this association, yet again then, is that the experience of dancing provides a person with the enabling conditions he needs in order to participate in the rhythms of his own self-overcoming. Those conditions include a knowledge of bodily becoming. While there may be other activities that fulfill the criteria for a physical–spiritual practice capable of enabling the rhythms of great health, it is not immediately evident what they might be. Only dancing (as Nietzsche envisions it throughout his corpus) is an *art* in which a person *enacts*—that is, both represents and participates in—the process of *bodily self-creation*.

Strengthening the Instincts

To have to fight the instincts—that is the formula of decadence: as long as life is ascending, happiness equals instinct.

—K6, 73; TI 479

Further indication of the kind of discipline and self knowledge Nietzsche's vision of dance involves appears in his discussions of the ingredients of great health, namely strong instincts and refined senses. As noted, modern men suffer their physiological contradiction because their instincts are weak and corrupted. As such, they have lost the ability to distinguish what promotes great health from what diminishes it: "The instincts are weakened. What one ought to shun is found attractive. One puts to one's lips what drives one yet faster into the abyss" (K6, 22; CW 165). As Nietzsche writes: "I call an animal, a species, or an individual corrupt when it loses its instincts, when it chooses, when it prefers, what is disadvantageous for it" (K6, 172; AC #6, 572). That the instincts can be corrupted and weakened again expresses Nietzsche's nuanced sense of the "natural." Instincts—such as the desires for freedom, self-preservation, and movement—may be "natural" but they are not by that account fixed or "given" in any determinate way. They are living, malleable. They can be fed or starved. They find expression in the ways we think, feel, and act. When they decay, they war against each other and we are too weak to endure the struggle. We seek to give our corrupted instincts free reign, or excise them all together. We seek escape from our own internal conflict, our physiological contradiction, into dream worlds of religion or science.

Given this predicament the solution—revaluation—cannot involve the use of mental powers to corral or control the instincts. Such a relation to oneself, as in the quotation above, *is* a "formula for decadence." Instead, Nietzsche urges readers to discipline themselves to their instincts—to learn to listen to their instincts, discern what they represent, and honor them as the sources of energy and knowledge that they are.

Nietzsche's longest discussion of what disciplining oneself to one's instincts involves appears on the pages of *Ecce Homo* when he describes his own physical–spiritual practice. He recounts what he has discovered about how best to eat, where to live, and what to do in his years of experimentation. As he notes, in his younger life, he ate and lived "unselfishly"; he consumed what was offered to him in both food and ideals (K6, 279–81; EH 237–9). He lived where his family was. He followed the path paved for him by the expectations of those he loved, and of those who loved him. He ordered himself according to the schedule and demands of a student, a scholar, and eventually a university professor. Then the sickness that had dogged him from youth grew intolerable. No longer able to live unselfishly, he was forced to attend to himself. His instinct for self-preservation demanded it. He sought out the conditions that would enable his convalescence.

Nietzsche's discussion in this regard harks back to the passage in Book 5 of *Science* where he describes dance as the ideal, art, and piety of a philosopher. There he offers the example of a dancer, in contrast to a scholar, as someone who knows how to eat: "It is not fat but the greatest possible suppleness and strength that a good dancer desires from his nourishment" (K3, 635; GS #381, 345–6). The fat is the tasty part, high in calories, low in nutrients. Though essential to good health, it is so only in small quantities. The scholar craves it. That a dancer does not and can eat for nourishment attests to her refined *instincts*. She has spent years experimenting with her experience to discern the conditions most conducive to her physiological health, her dancing self. She eats for nourishment not because she is able to use her mental will power to corral her desires, but because she has learned through her own practice and experience to heed and honor the rhythms of her bodily becoming. As a *result* of her practice, she desires *instinctively* to eat what promotes her physiological well-being, her ability to create value. As Nietzsche says of the free spirit in *Human*, her virtues are positive, not negative. Dancing serves as an ideal, art, piety, insofar, as dancing is an *activity* in which a person cannot not discipline her consciousness to the bodily conditions of her great health.

Further, dancing is a kind of activity that not only guides a person in attending to his instincts, but strengthens those instincts. As the instincts grow stronger so too does the conflict among them. However, a person with strong instincts also develops the ability to tolerate the conflict, and even welcome that conflict, as the condition out of which he can become who he is. As Nietzsche remarks, the instincts "contradict, disturb, destroy each other"; and as some grow and others wither, the shape of an individual emerges, as a *whole* (K6, 143; TI 546). To be whole, *ganz*, in this sense is to embrace one's instincts as the materials out of which we create ourselves and as the energy of our creative impulse. For this reason, it is only when we can develop the strength to sustain our own warring multiplicity that we can fully unfold our power to create values. Nietzsche's model in this regard is Goethe: "he did not retire from life but put himself into the midst of it . . . What he wanted was *totality*; he fought the mutual extraneousness of reason, senses, feeling, and will . . . he disciplined himself to wholeness, he *created* himself . . . he said Yes to everything" (K6, 151; TI 554). To discipline oneself to wholeness is to enact a love for one's self so great that it expands to include, forgive, and redeem all dimensions of oneself. As Nietzsche confirms, "The price of fruitfulness is to be rich in internal opposition; one remains young only as long as the soul does not stretch itself and desire peace" (K6, 84; TI 488). At the same time, this embrace of internal opposition is not static; self-love is not self-indulgence. To love oneself in this way is to unleash one's greatest potential for creating oneself anew—for what Nietzsche calls the "spiritualization [*Vergeistigung*]" of the instincts and desires. To spiritualize the instincts is not to subordinate them to a "spirit." It is rather to acknowledge the life or spirit moving in them; it is to honor the wisdom they represent—a wisdom that may take years of discipline and obedience to discern. It is to create one's gods and virtues by engaging in the practice of honoring the tensions and rhythms of bodily life.

That dance may be helpful in negotiating such enabling conflict among the instincts is suggested in a number of places reaching back to *Birth*. As discussed, Nietzsche almost always depicts dancing as a mediating movement, arcing between Dionysian and Apollinian, between intuition and reason, between religion and art on the one hand and science on the other. The act of dancing effects this mediating moment in every case, and does so by facilitating a physiological transformation. A person comes to experience herself differently—as the movement of her own bodily becoming, as a kinetic image maker. Dance enables her to relearn her relationship to the "small things" that are "inconceivably more important than everything one has taken to be important so far." As Nietzsche confirms, "precisely here" one must begin to relearn [*umzulernen*]" (K6, 294; EH 256). In this regard, the pieces are in place: "dancing" stands poised as an ideal of the kind of practice we need in order to experience the warring of our instincts as the condition for exercising our freedom in creating and becoming our highest ideals—even near abysses.

Refining Sense

"Reason" is the cause of our falsification of the testimony of the senses. Insofar as the senses show becoming, passing away, and change, they do not lie.

—K6, 75; TI 480–1

As the instincts grow stronger and more "spiritual," Nietzsche explains, not only do we become more capable of discerning what is good for us and thus of becoming who we are, our senses develop a subtlety they did not before enjoy. We become more attuned to smaller differences in the choices we make concerning food, climate, and recreation, our goals, values, and ideals. Where adherence to the ascetic ideal guides a person "to draw in his senses, turtle fashion, to cease all intercourse with earthly things, to shed his mortal shroud" and thus become "perfect," Nietzsche counters: "[I]f we subtract the nervous system and the sense—the 'mortal shroud'—*then we miscalculate*" (K6, 181; AC #14, 581). Active, vital senses are necessary to the process of (self) knowledge and not just because they are the media through which "we" experience the "world." Our senses are themselves creative (as noted in chapter 1); they are the means though which we generate and become images of our selves.

In this regard, Nietzsche discusses art at length in his later works and his paradigm for art, as in *Human*, bears traces of his account of dance in Attic tragedy. Where the primary responsibility of art is to educate and refine the senses, art does so, in this account, by stimulating our instinct for gestural imitation. For example, Nietzsche writes: "All art exercises the power of suggestion over the muscles and senses, which in the artistic temperament are originally active: it always speaks only to artists—it speaks to this kind of a subtle flexibility of the body" (WP #809, 427). Art stimulates our kinetic response; we move our eyes, ears, and muscles, and by moving notice differences, small gradations in color, shape, sound, movement. In so far as art stirs movement responses in us, it provides us with an occasion to reflect on our sensory engagement with the world. We not only exercise our senses, then, we sense how we are becoming through our own acts of sensing. Art awakens in us the primary level of metaphor-making. In so doing, art generates in us the vitality we need in order to endure the conflicting messages of our senses, and learn to discern the conditions for our ultimate satisfaction. Art, then, in so far as it not only stirs the senses, but provides people with opportunities to exercise them, facilitates the process of *Vergeistigung* described above.

Finally, in offering this dance inspired account of art, Nietzsche rejects the idea that art creation or appreciation is the privilege of a few. All one needs in order to dance is what every person already is: a body. Moreover, all persons, as bodies, are inherently creative. Thus, the artist is a person who has not forgotten his bodily becoming, and who, in making art, speaks to the artist in each of us. As Nietzsche writes, echoing his earlier theories of language: "our faculties are subtilized out of more complete faculties. But even today one still hears with one's muscles, one even reads with one's muscles" (WP #809, 428). That one still "hears with one's muscles" (as Nietzsche describes as well of therapeutic Greek music in *Human*) is evidence of both a human instinct for movement and a capacity for making and perceiving kinetic images. That one still hears with one's muscles in turn suggests that we are not too far gone to develop an appreciation of the dance in art, and in fact, of the dance in all communication. As Nietzsche confirms: "One never communicates thoughts: one communicates movements, mimic signs, which we then trace back to thoughts" (WP #809, 428). It is only when our instincts are weak and our senses dull that we fail to appreciate dance as (an effective symbol of) theopraxis.

Conversely, once there are people who can develop forms of art that realize this vision for dance, it follows, more of us will regain the ability to sense and appreciate not only our need for dance, but the enabling role that dancing can and should play in our artistic, religious, and scientific creations. If we forget the "mortal shroud" as Nietzsche suggests we do, we fall prey to religious fanaticism: "a sort of hypnotism of the whole system of the senses and the intellect for the benefit of an excessive nourishment (hypertrophy) of a single point of view and feeling that henceforth becomes dominant— which the Christian calls his *faith*" (K3, 582; GS #347, 289–90). If we act out of "fear of *over-powerful* senses," we generate philosophical idealisms that displace the meaning and value of our bodily existence to an other, abstract world (K3, 624; GS #372, 333). In either case, we perpetuate ignorance of ourselves. By contrast, if we discipline ourselves to our instincts, and exercise and refine our senses, we engage our value-creating, metaphor-making potential in overcoming ourselves, becoming who we are, and developing great health. We revalue self-sacrificing love by incarnating an alternative: an overflowing, self-enjoying love. As Nietzsche insists, the "spiritualization of sensuality is called *love*; it represents a great triumph over Christianity" (K6, 84; TI 488).

Stirring Energy

> *We others, we immoralists, have, conversely, made room in our hearts for every kind of understanding, comprehending, and* approving. *We do not easily negate; we make it a point of honor to be* affirmers.
>
> —K6, 87; TI 491–2

In facilitating self-knowledge, strengthening our instincts, and refining our senses, the dancing Nietzsche envisions provides us with the very conditions we need in order to be free spirits. Dancing builds *energy* in us. It builds the sense of vitality and power that provides us with a perspective from which to reflect critically and creatively on all aspects of culture. Dancing develops in us the energy needed to "make room in our hearts for every kind of understanding"—in short, the energy needed to love.

Some of the wisest discussions in *Ecce Homo* concern Nietzsche's recommendations on how to manage our energy. He counsels us not to squander energy in postures of revenge or defensiveness; to avoid situations in which we constantly have to say "no." As he elaborates, our instinct of self-preservation,

> commands us not only to say No when Yes would be 'selfless' but also to say *No as rarely as possible*. To detach oneself, to separate oneself from anything that would make it necessary to keep saying No. The reason in this is that when defensive expenditures, be they ever so small, become the rule and a habit, they entail an extraordinary and entirely superfluous impoverishment. Our *great* expenses are composed of the most frequent small ones . . . Merely through the constant need to ward off, one can become weak enough to be unable to defend oneself any longer. (K6, 292; EH 252)

The effect of such holy selfishness, ironically enough, looks like what Christianity (also) teaches: tolerance, forgiveness, gratitude, and love.

That dancing can serve this function of teaching us how to navigate the currents of energy coursing through us is evident in Nietzsche's discussions of art in *Case* and *Twilight*. In these works, Nietzsche's criticism of Wagner's music echoes one he began in *Science*: Wagner's music fails to make Nietzsche dance. He writes, "My objections to the music of Wagner are physiological objections . . . My 'fact' is that I no longer breathe easily once this music begins to affect me; that my foot soon resents it and rebels; my foot feels the need for rhythm, dance march; it demands of music first of all those delights which are found in *good* walking, striding, leaping and dancing. . . . What is it that my whole body really expects of music? I believe, its own *ease*: as if all animal functions should be quickened by easy, bold, exuberant, self-assured rhythms" (K3, 616–17; GS #368, 324–5). As noted, Nietzsche contends that Wagner's music weakens the instincts, corrupts the senses, and depletes our vital energies by overexciting our nervous systems. Dragged through a dizzying array of emotional states, we lose consciousness of how exhausted we are (K6, 22–3; CW 165–6). What we miss in his music, Nietzsche elaborates, is the "la *gaya science*; light feet, wit, fire, grace; the great logic; the dance of the stars; the exuberant spirituality; the southern shivers of light; the *smooth* sea—perfection" (K6, 37; CW 178). In this perplexingly poetic list, the metaphors all refer back to moments in *Science* or *Zarathustra* where Nietzsche describes the process of overcoming oneself as one of taking up the responsibility of being a creator of value.

In *Twilight*, Nietzsche expounds on the alternative that a music which dances would represent. It would transform us, or occasion a transformation in us, by inciting "frenzy." Frenzy is "the feeling of increased strength and fullness" that impels persons to impose forms on what appears to them—to idealize; it is the feeling that fuels a "tremendous drive to bring out the main features so that the others disappear in the process" (K6, 116; TI 518). This feeling of strength overflows into experience such that everything a person experiences "is seen swelled, taut, strong, overloaded with strength," a "mirror" of his power, "reflections of his perfection" (K6, 116–7; TI 518–9). While admitting that it comes in Apollinian and Dionysian forms, Nietzsche suggests that it is the Dionysian frenzy, a frenzy of the body, that we who suffer from a physiological contradiction risk losing. We have "immobilized" our senses, "especially the muscle sense" such that we are unable to sense whether the music makes us dance or not (K6, 118–9; TI 520). In this state we are incapable of sensing and responding to our own drive to move. We have lost the ability to sense that moment in which "the whole affective system is excited and enhanced: so that it discharges all its means of expression at once and drives forth simultaneously the power of representation, imitation, transfiguration, transformation, and every kind of mimicking and acting" (K6, 117; TI 519). When a person does allow himself to hear music through his bodily movement, he "enters into any skin, into any affect: he constantly transforms himself" (K6, 118; TI 520). As people lose the ability to listen through their bodies, then, they risk losing the ability not only to discern whether or not the music is enhancing or depleting their energy stores, they lose the lived knowledge of themselves as metaphor-makers that is possible through that experience as well as all the fruits that come with such practice.

To conclude this discussion, several points are worth noting. First, what constitutes the senses as with the instincts is a function of how we live in the world, of the disciplines we engage or not, of what we believe. As such, when Nietzsche defines an instinctual "drive" to move and a "muscular sense," he is engaged in the process of revaluing Christian values. As he insists, Christian values have "taught men to misunderstand the body" (K6, 231; AC #51, 633). That misunderstanding consists in understanding the body as a thing rather than a process of its own becoming. Nietzsche's revaluing response is not to offer a new concept of what a body actually is, but to discern the mechanisms, the practices, by which a body becomes what it claims to be. In other words he reads our conception of and relationship to "the body" as a symptom of our physiological health. In this way, he both affirms the dichotomy between body and mind as a valid expression of a physiological state while providing a perspective for critiquing that dichotomy as expressing an impoverished state. In this way, Nietzsche enacts his love for Christian values as enabling him to imagine a better alternative.

Second, that alternative, time and again, involves *dance*. In Nietzsche's account, dance emerges as an *ideal* of the art best positioned to help people to a physical consciousness of their bodily becoming. Yet it serves as an *ideal* because it represents a practice: the act of dancing cultivates attention to the instincts (including the instinct to move), refines the senses (including the muscular sense), and stirs and communicates energy needed to tolerate the ambiguity of one's own processes of sensory metaphor-making. Dance as Nietzsche envisions it, provides persons with a way of tapping the frenzy—the levels of affective animation—needed to awaken and animate their creative ability. In all these respects, dancing counters the effects of living in accord with the ascetic ideal. It does so, not by attacking the ascetic ideal at the level of rational argument, but by representing a discipline of self-knowledge that provides persons with a physical consciousness whose reality ascetic ideals deny. In teaching us to tap and channel our energies, dancing trains us in the rhythms of (self) love. For as Nietzsche writes, it is "Love" that "gives the greatest feeling of power . . . [and man] calls his feeling of love God" (WP #176, 107). Consequently, what we need in order to revalue the ideal of self-sacrificing Christian love, rather than music that allows us to listen passively, is music that impels us to dance, spaces that allow us to dance, and forms of dance that educate us in forms of bodily frenzy. In this way, the *action* of dancing emerges as a criterion for the conditions necessary for great health.

To dance is to love. To love is to dance. To dance is to make the movements of Zarathustra in his movements toward humanity. To dance is to overcome Christian morality by loving what it teaches us about how to overcome our sickness. To dance is to incarnate the revaluation of Christian values—to enact the counter to the ascetic ideal, by engaging in a discipline of bodily self-knowledge. Dance is (an ideal of) theopraxis: an activity in which we make and become our images of god.

Against Writing

In mounting a radical challenge to a culture predicated on the ascetic ideal, Nietzsche's ideal for dance not only resists the forces of desensualization and oversensitization, it also counters the "faith in grammar" that drives the textualization of culture. As Nietzsche avers, "I am afraid we are not rid of God because we still have faith in grammar" (K6, 78; TI 483). I have already noted how Nietzsche recommends that people learn to think and write as a kind of dancing, and that he insists that dancing should be a part of a noble education. In these later works, his critique extends even further. Writing, he claims, kills. What can be written is already dead and gone. As such, language is limited in its ability to help us know ourselves as individuals. "We no longer esteem ourselves sufficiently when we communicate ourselves. Our true experiences are not at all garrulous. They could not communicate themselves if they tried. That is because they lack the right word. Whatever we have words for, that we have already got beyond" (K6, 128; TI 530–1).

Moreover, reading and writing fail not only in providing us with self-knowledge in general. They fail to provide us with the particular kind of self-knowledge we need in order to embrace, navigate, and overcome the physiological contradiction we suffer as modern men. What we need are practices like dance that not only strengthen us physiologically, but do so in ways that provide us with the energy, the clarity of heart and mind, to listen to all contradictory aspects of ourselves and find our health through them. As Nietzsche suggests time and again, in order to learn to write, we need to do other kinds of activities than just write. We need to develop a perspective within ourselves other than that opened by writing in order to think about writing and think through writing in life-affirming ways. Most of all, we need not to sit. We need to walk; we need to dance. As he confirms, "The sedentary life is the very sin against the Holy Spirit. Only thoughts reached by walking have value" (K6, 64; TI 471). That Nietzsche claims that "God is dead" does not invalidate his point. His Holy Spirit is a god who could dance. The point is that the act of sitting does not provide us with the experiences we need in order to generate gods who can dance. Sitting thoughts afford an illusion of abstraction; while sitting we are very good at imagining that we have no body. Thoughts generated while walking, at least for Nietzsche, can express the relation to embodiment characterized by attentive, loving respect that impels him to walk in the first place. Free movement is the source of life-affirming thought.

Thus, when Nietzsche confirms: "Sit as little as possible; give no credence to any thought that was not born outdoors while one moved about freely—in which the muscles are not celebrating a feast, too. All prejudices come from the intestines" (K6, 281; EH 239–40), he is serious in his laughter. While it is tempting to read such passages as poetic hyperbole, the thrust of Nietzsche's physiological method and his critiques of sitting to write encourage otherwise. Writing immures us in edifices of our own construction unless we consciously and constantly stop writing, and empty ourselves in an awareness of bodily movement, feeling our senses, our muscles, our beating, pulsing, rhythmic selves, as the source of our becoming. What we think will always express our bodily being, even and especially when we deny that to be the case.

Mourning Sickness

Though hast turned for me my mourning into dancing.

—Psalm 30

Nietzsche, as noted in chapter 1, did not study dance. He did not engage in a regular practice of dance. As such there is a sense in which his choice of dancing as a metaphor for the practice of revaluation still seems perplexing. However his repeated references to the dance epidemics of Christian history offer a final clue. By Nietzsche's own account, his sickness provided him with the educating role he projects onto images of dance. Nietzsche's sickness prevented him from conforming to the expectations placed upon him by his family, religion, profession, or society. It prevented him from acting unselfishly. It forced him into an acute self-knowledge of *how* his various habits, scholarly practices, and philosophical and religious ideas affected his bodily health: ". . . my sickness . . . forced me to see reason, to reflect on reason in reality" (K6, 282; EH 241). It forced him to see the limitations of "This 'education' which teaches one from the start to ignore *realities* and to pursue so-called 'ideal' goals" (K6, 279; EH 237). It drove him to look to refine his senses, rely on his instincts, and conserve his energy in the face of his struggle. Nietzsche's sickness, in other words, forced him to develop a physical consciousness—a critical perspective within himself from which to assess and revalue the value of his values and ideals.

Moreover, it encouraged him to discover and exercise his capacity for generating and becoming kinetic images of his highest ideals. In the space opened up by his inability to attach to people and places, Nietzsche describes how his "nethermost self" awakened (K6, 326; EH 287–8); he lived experimentally. He exercised his bodily creativity in sensing and responding to his own sickness, in determining its meaning for his "self." He found in that sickness, wisdom—guidance in discerning and overcoming the decadence of modern culture. As he relates in *Ecce Homo*, now "I take readings from myself as from a very subtle and reliable instrument" (K6, 282; EH 241). In short, Nietzsche became healthier (happier, freer, more responsive) in his sickness than those whose health enables them to conform to what is expected of them. He developed a love for his bodily becoming—for his sickness and his health—so strong that it overflowed to include all of life. And for this education to bodily becoming, Nietzsche expressed thanks: "Yes, at the very bottom of my soul I feel grateful to all my misery and bouts of sickness and everything about me that is imperfect, because this sort of thing leaves me with a hundred backdoors through which I can escape from enduring habits" (K3, 536; GS #295, 237)— including the "enduring habit" of believing in God.

Yet the lesson he takes from this dynamic is not that we should all contract some mysterious illness in order to achieve a health-enabling perspective on life. In his opinion, we are already sick. Further, our participation in making ourselves sick is precisely the source of our greatest suffering. Instead, Nietzsche distills his lesson differently: people need some kind of physical–spiritual practice, some way of working in and with their bodies, through which they come to know the great reason, healing power, and infinite wisdom of bodily becoming. It is in this sense, then, that Nietzsche's sickness provided him with a perspective and a will to appreciate the images of dancing with which he was familiar not just as signs of ecstatic self-loss, or overflowing health but as representing a practice of recreating or overcoming oneself.

If Nietzsche had not been sick, he may never have developed dancing as a dominant metaphor for the process of overcoming oneself.[5] Because of his sickness Nietzsche learned how he and not God would turn for him his "mourning into dancing." Nietzsche turned his mourning into dancing by mounting a vision of dance not yet realized in the forms of modern culture: a vision of dance as theopraxis—as an activity in which we incarnate an alternative to self-sacrificing love through our willingness to participate consciously in the rhythms of our bodily becoming.

Part 2

Isadora Duncan

The stage is suffused with light—amber, violet, and rose. The glow reflects off blue–gray curtains and rug. Members of the orchestra shuffle into silence. Their music begins, swelling, soaring.[1]

A lone dancer appears—Isadora Duncan. She is luminous, draped in fabric, flowing and sheer. She seems to float, as if propelled by a strong current. Her legs churn, pressing firmly off the floor and returning softly, as she is twirling, skipping, reaching, and falling. Her movements evoke waves—the tide on the beach, the curl and crash of a rising force, ebb and flow. Forward and back, across the stage, right to left and left to right, cycles of movement, turning and returning, sweep the spectator into a world washed with enabling power.

Her dance is *Blue Danube* (1902), a popular favorite. Yet there is no river on stage. Nor is the dancer pretending to play in an imaginary stream. There is no other river in the piece than that evoked by the movements of her graceful, flowing body.

Rather the title invites us to see water anew—to see a river in a dancer's movements, as represented in those movements, as capable of being represented in those movements. It invites us to see how a person can recreate in her own movements the movement shapes or kinetic images in which a river appears to her. She can evoke a sense of river, a sense of human being as related to river, a sense of Nature as an ordered pulsing whole including human, animal, and natural worlds, and a sense of a human participating in these currents of creation as a dancing body.

As the dancer flows and pulses, skimming and dipping across the stage, the force of her movements arrests our attention. We are drawn in to her intoxicated, intoxicating reverie; to her discipline of attending to the power coursing through her healthy body; to the workings of her kinetic imagination. We identify with that power and health. Its ecstatic dimension is contagious. We know ourselves to be capable of the kind of experience—the harmonious, enabling relation to nature—her dancing represents. We swell and surge along with her, witnesses to the sense of self and world her trained body enables us to comprehend. We know ourselves as bodily becoming, as ever engaged in the process of creating and incarnating our highest ideals.

Chapter 4

A Dionysian Artist

I had come to Europe to bring about a great renaissance of religion through the Dance, to bring the knowledge of the Beauty and Holiness of the human body through its expression of movements, and not to dance for the amusement of overfed Bourgeoisie after dinner.

—ML 85

The dance is not a diversion but a religion, an expression of life.

—AD 142

Isadora Duncan was concerned. Modern culture was in the throes of a crisis. Its forms of religion and art, Christian in particular, were failing to provide people, especially women, with the inspiration or guidance they needed in order to honor their own bodies. In the quest for enlightenment of various kinds, people were educating themselves to ignore their bodily being and to abide instead by the stern power of their brains, imposing codes of morality onto themselves. She was aware of the cost: "[V]ery soon the movement is imposed from without by wrong theories of education, and the child soon loses its natural spontaneous life, and its power of expressing that in movement" (AD 77). Duncan, raised to think of religion as a human creation, responded with a call to reinvent religion: nothing less than a "great renaissance" or rebirth of values was necessary to help people appreciate and love *human bodies* as partaking in what she perceived as the highest values of humankind: Beauty and Holiness.[1]

Further, given the direction of change needed, Duncan insisted that *dancing* must serve as the conduit of such a renaissance. The knowledge required to overcome the "paralysis of bodily expression" (AD 77) must come through a body's own "expression of movements." Duncan embraced dance, the art whose medium is bodily movement, as providing a unique opportunity for humans to come to know their human bodies as source and site for creating values, including values of beauty and holiness. Dancing, she believed, could provide a model for what religion should be—an "expression of life."

In calling for a rebirth of religion through dance, however, Duncan was not naïve. As she readily admitted, the forms of dance capable of bringing about such a renaissance did not exist. Its popular and artistic forms both embodied the attitudes toward bodily being ensconced in Christian values. In classical ballet, for example, the favorite of the "bourgeoisie," Duncan observed how dancers exercised the same relation to their bodily selves she criticized in Christian morality—they imposed formal codes onto their bodies in order to project images of the body as weightless, ethereal, offering little resistance to mental direction. Such forms of dance, she argued, reinforced perceptions of dance as a physical activity, as mere amusement.

Thus, Duncan's vision for dance was a double vision. Her call for a "dancer of the future" was also a call for a renewed religion. In turn, the rebirth of religious values she desired awaited the creation of new aesthetic values—forms of dance technique, training, choreography, and performance—through which people could come to know and value their bodies in a different manner than they had learned from Christian morals.

While Duncan describes having had her double vision before reading Nietzsche, nearly all her descriptions of that vision appear in the years after she had read Nietzsche. Upon reading his work, Duncan fused her vision with his.[2] In the essays from 1902–1903 to 1927 she collected for publication in what become the last year of her life, nearly every one of the pieces alludes to Nietzsche. She describes him as a "great Master" of the dance; as a philosopher who understands the "spirit" of the dance. She paraphrases his arguments, especially those from *Birth of Tragedy*; she cites him directly, most often pulling quotations from *Zarathustra*, and she picks up and employs his vocabulary in describing her dance process. She writes about Dionysian energies, free spirits, and healing art. In fact, his name appears more frequently in her writings than any of her other "teachers," a list that includes Ludwig Beethoven, Walt Whitman, Richard Wagner, Charles Darwin, Ernest Haeckel, and Jean Jacques Rousseau. She chose a quotation from Zarathustra's "Yes and Amen Song" for the frontispiece of her autobiography.[3]

Why Nietzsche? As this chapter illustrates, Duncan perceived in Nietzsche someone who shared her double vision: a vision for what dance can and should be, namely "religion"; and a vision for what religion can and should be, namely an "expression of life." As she writes, "No art antique or modern has been able to reveal all that man can be when inspired by his highest aspirations, in terms of movement . . . Nietzsche had a vision of it" (AD 119). When she writes that *Zarathustra* "is filled with phrases about man in his dancing being" (AD 123), she acknowledges that Nietzsche, like her, perceived humans *as* dancing beings and dancing as integral to the process of fulfilling our humanity. When she insists that Nietzsche was not speaking of the "execution of pirouettes" but the "exaltation of life" (AD 77), she confirms that for her, as for him, dancing is a theopraxis—an activity in which people image, elevate, and sanctify their highest ideals, and thereby enact the process of becoming or self-overcoming that Nietzsche claims "life" is. Moreover, when Duncan calls for dance that (is religion and thus) expresses life, she uses the term "expression" not to reify a split between internal and external but in a Nietzschean sense to characterize a process of bodily self-overcoming. For Duncan, dance is an expression of life the way fruit is an expression of a tree. *Dance enacts the potential of life to create beyond itself.*

In sum, Duncan saw in Nietzsche someone who comprehended her vision of dance as capable of generating and realizing an alternative morality to the one offered by Christian sources—an alternative that affirms life, health, sensuality, illusion, and bodily becoming. In turn, she imagined herself a dancer-philosopher. She would develop principles of practice, choreography, and performance capable of catalyzing this mutual renaissance of religion and dance. She believed she had. At one point in her autobiography, after quoting from "On the Higher Man" in the Fourth Part of *Thus Spoke Zarathustra*, Duncan comments on a 1914 performance of her students. She crows, "These were indeed the gestures of the Vision of Nietzsche" (ML 301, with a quotation from Z 406).[4]

Given this history, it is impossible to isolate Nietzsche's influence on Duncan's evolving vision for dance and religion. So deeply was she influenced by his writing that time and again his account of revaluing values sheds light on the claims she makes for her dances even when she is not directly referencing him.[5] This chapter, then, traces the religion language Duncan uses in describing her vision for dance in order to demonstrate the kinship between her project of renewing religion and Nietzsche's of revaluing values. This discussion sets the stage for chapter 5, where I present Duncan's practice, choreography, and performance as an illuminating case study in how to overcome Christian morality, revalue values, and conduct a radical affirmation of life.

Engaging Christianity

*I was probably never much more advanced than to be a Pagan Puritan, or a
Puritanical Pagan.*

—ML 255

The strongest evidence of the degree to which Duncan allied her vision with
Nietzsche's appears not in explicit claims that her dances realize
Zarathustra's vision, but in the ways she describes her relationship to
Christianity in particular and religion in general. Duncan repeatedly
demonstrates an ambivalence with respect to Christianity, one for which
people tend to criticize her as muddled, inconsistent, and unsystematic,
a dancer and not a philosopher. However, in light of Nietzsche's work, this
ambivalence is an indication that Duncan located her work squarely in the
modern predicament Nietzsche describes. She perceived herself, in her
passion for dance, as embodying a *physiological contradiction*—she was the
site where competing modes of valuation that she identifies as "Puritan" and
"Pagan" were at war with one another. In calling for a dancer of the future—
"the free spirit, who will inhabit the body of new woman; more glorious
than any woman that has yet been . . . the highest intelligence in the freest
body!" (AD 63)—Duncan aspires to do what Nietzsche's Zarathustra
teaches: to transform her experience of physiological contradiction into
a stimulus for making art that revalues religion. From this perspective,
Duncan's ambivalent references to Christianity distinguish her as one of
Nietzsche's free spirits: in her dance Christian religion is overcoming itself.

At the same time, in her use of the terms "Puritanical Pagan" and "Pagan
Puritan," Duncan invokes slightly different terms of physiological contradiction than those articulated by Nietzsche. For Duncan, this contradiction
hinges not only on Christian attitudes toward bodily becoming, but also on
Christian attitudes toward women and female bodies in particular. On the
one hand, Duncan strongly denounces her "Puritan" heritage, rejecting
the teachings and practices of the Christian church for the way they seduce
women into positions of submission vis-à-vis men. In this regard, Duncan
was strongly influenced by her mother, Mary, who renounced her own
Catholicism just after Duncan, her fourth child, was baptized at the age of
five months. Mary then divorced a husband she deemed unreliable and pro-
ceeded to raise her children on her earnings as a piano teacher. The children
lived the repercussions of Mary's choice daily. Always short of funds, moving
from home to home, they found comfort and inspiration in Mary's love of the
arts, and her readings to them from literature, including lectures by atheist

Robert Ingersoll. The children studied together, wrote dramas together, and formed a family theater troupe. By the time Isadora was 16, the family was touring the California coast, with Mother playing the piano.

Following in Mary's footsteps, then, Duncan decried Christian teachings and practices that encourage hostility toward dance, human bodies, or the bodies of women. In lectures and essays she railed time and again against gender stereotypes, social practices, styles of fashion, and theories of education she perceived as expressing Christian antipathy to (women's) bodies: "From my earliest childhood I have always felt a great antipathy for anything connected with churches or Church dogma. The readings of Ingersoll and Darwin, and Pagan philosophy had strengthened this antipathy" (ML 277). Eventually it was a Nietzschean "spirit," Duncan suggests, that enabled her to move out of the "prisonhouse of the days before the Emancipation of Women" where her mother "could only suffer and weep." As she describes: "This is the Christian education which does not know how to teach children Nietzsche's superb phrase: 'Be hard!' Only from an early age some spirit kept whispering to me to 'be hard' " (IS 28).[6]

Yet, in moving out of the "prisonhouse" Duncan did not reject Christian religion completely, even in its Puritan manifestations. She identifies herself as a "Puritan"; she admits how deeply steeped her physical experience is in a Christian sensibility. For Duncan, as for others in her day, "Puritan" represented a mode of valuation that encouraged people to ignore their senses and instincts as having any role to play in what is most important— a relationship to God.[7] A Puritan is an idealist, a bound spirit. As Duncan elaborates, the goal embedded in the "American trend of education is to reduce the senses almost to nil. The real American is not a gold chaser or money lover, as the legend classes him, but an idealist and a mystic" (ML 79). Yet Duncan acknowledges the influence of this education on her development as a dancer. She goes so far as to describe her younger self as "a product of American Puritanism": "the land of America had fashioned me as it does most of its youth,—a Puritan, a mystic and a striver after the heroic expression rather than any sensual expression whatever" (ML 78). It was the Puritan in her, able and willing to "reduce" her senses, who yearned for transcendence and invented "gods" of "Beauty and Love" for herself (Terry 1963: 91). Thus, Duncan was aware, as Nietzsche had been, that her search for truth through dance expressed a faith in the ascetic ideal she had imbibed from her Christian heritage. Duncan's ambivalence, in this regard, echoes Nietzsche's insight: the ascetic ideal is the form in which Christian morality is overcoming itself.

Given her awareness of how this contradiction lives within her flesh, Duncan was savvy enough to know that she could not simply shrug it off by refuting Christian values. Nor would she want to. Instead, she sought to engage the creative power ascetic ideals represent—human beings' god-forming instinct at work—and exercise that power in creating ideals that encourage people to train their senses differently, ideals that remain faithful to the earth. She called upon dance to provide women in particular with a means for liberating themselves from the ubiquitous hostility toward embodiment she named "Puritan." As she insists: "If my art is symbolic of any one thing, it is symbolic of the freedom of woman and her emancipation from the hidebound conventions that are the warp and woof of New England Puritanism" (IS 48).

From this perspective, Duncan's ambivalent statements regarding "Puritan" Christianity signal her participation in Nietzsche's project of revaluing values. Her love for *dancing* expresses *both* her Puritan heritage *and* her opposition to it; and it does so in such a way as to use the tension of this physiological contradiction to impel her in creating new forms of dance—forms that embody new religious values. She, like Nietzsche, seeks to transform her experience of this contradiction into a stimulus for loving life, for making the movements of Zarathustra's dance. At the same time, her formulation of that physiological contradiction also discloses how she critically advances Nietzsche's project along the lines of her commitment to it. For Duncan, any revaluation of Christian morality will not be complete until *women dance* their *religion*. Duncan's vision, in other words, is three-fold: dance will realize its potency as art when dancers reinvent how "religion," namely Christianity, values "woman." Describing her vision, Duncan admits that it was the same "dream that had resounded through the words of Christ" (ML 359).

Going Greek

[T]he dance of the future will have to become again a high religious art as it was with the Greeks. For art which is not religious is not art, is mere merchandise.

—AD 62

Further evidence that Duncan's triple vision for dance engages and advances Nietzsche's project of revaluing Christian values appears in her discussions of the ideal that sustains her resistance to Christianity, namely the "Pagan" world of ancient Greek culture and philosophy. While Duncan nourished an affinity for ancient Greek culture long before reading Nietzsche, she came to articulate her relationship to ancient Greece in terms akin to his. Neither Duncan nor Nietzsche advocated a return to the time of ancient

Greece; neither (after an initial failure on Duncan's part) aimed to recreate
the forms of Greek culture. Rather, both embraced ancient Greece as an
ideal—an ideal of an alternative mode of valuation, that identifies and honors what Nietzsche and Duncan after him called the *Dionysian* energies of
life. Yet again, in her allusions to the Greek god, Duncan finds justification
for why the revaluation of religion through dance (and vice versa) requires
that women dance.

In admiring the worldview of ancient Greek culture as posing an alternative to that shared by modern western bourgeoisie, Duncan and
Nietzsche were not alone. In the last decades of the nineteenth century,
everything Greek was the rage. Accounts of archeological discoveries by
Heinrich Schliemann and others made the newspapers. Schliemann settled
in California, and by the time he died in 1890, American culture was
saturated with replicas of Greek art, stories of Greek myths, and Greek-style
techniques for draping fabrics and clothing (Sorrell 1981:321). Duncan
herself attributes her early interest in Greek culture to a reproduction
of Botticelli's "Primavera" that hung above her bed. Its image of Venus
rising from the waters guided her choice of dance costumes for years.
Further, the romantic poets, artists, and musicians her mother adored had
also idealized Greece. From early on, Duncan thus appreciated "Greece" as
an ideal that inspired her in her attempt to use dance to renew Christian
religion along the trajectory of her commitment to it. As Duncan wryly
confirms: "If I am Greek, it is the Hellenism of Keats' 'Ode to a Grecian
Urn' " (IS 47).

Where Nietzsche had entered ancient Greece through literary and philosophical texts, Duncan entered through the study of its vases, friezes, and
sculptures. When she and her family decided to travel to Europe in 1899,
eager to find audiences for Isadora's art, they were even more thrilled, perhaps, with the prospect of seeing for themselves the Greek art housed in the
museums of London, Paris, and Munich. The British Museum, with its
Elgin Marbles, was one of their first stops. Isadora and her brother
Raymond spent hours in its galleries. There Isadora saw art peopled with
beautiful bodies—moving, dancing bodies, lightly-clothed with draped
fabrics that accentuated the lines of the body in movement. She saw women
dancing freely as goddesses, priestesses, and devotees. Noting the careful
attention given to these figures, Duncan concluded that the Greeks
accorded a value to bodily movement not matched in Christian culture: in
this art *dancers* not only appeared as beautiful, they appeared as *holy*, as gods
and goddesses, kinetic images of the Greek's highest, religious ideals. While
Raymond sketched, Isadora arranged her body in the poses she observed,
seeking to discover how to make movements that would communicate, as
she thought these figures did, the beauty and holiness of a human body.

In 1902 with receipts from Isadora's first commercial successes, the Duncans traveled to Greece and purchased land. At this time, they did try, unsuccessfully, to recover and restage ancient forms of Greek drama, music, and dance. As their funds dwindled, Duncan returned to Europe for further touring. It was in that fall of 1903 in Munich that she began her study of Nietzsche in earnest. Whether or not she was initially drawn to Nietzsche by their common interest in ancient Greece, it is clear that in the wake of her failed attempt to recreate Greek drama, Duncan found in him an alternate route for translating her attraction to Greek culture into forms of dance capable of catalyzing a rebirth of religion. She would create dances that occasion an enabling encounter with Dionysian wisdom. While she may have read other books by Nietzsche, his "aesthetic" works, *Birth of Tragedy* and *Zarathustra,* remained her favorites.[8] Closer examination of how Duncan uses the term "Dionysian" in describing the kind of experience she wanted her dancing to convey suggests how important Nietzsche's philosophy was in helping her craft a vision for a dance that would do for her time what she perceived Greek dancing did for theirs, namely, catalyze an affirmation of life.

Making Waves

It was to revive the lost art of dancing that I have devoted my life.

—ML 220

In her study of Greek artifacts, as she sought to understand for herself what the Greek artists knew about dance, Duncan made a discovery that proved seminal in the development of her vision for a religion-renewing, woman-liberating dance: the *undulating line.* She writes, "In the thousands and thousands of figures which I have studied on these vases, I have always found an undulating line as the point of departure" (AD 90). Due to this undulating line, figures fixed in stone or clay nonetheless project an illusion of movement. A raised foot arcs back toward the ground; a lifting arm is posed to fall back again; a head tilts backward, following an irresistible impulse through the body, and prepares to double back. As such the figures seem to convey what Duncan calls, with allusions to Nietzsche, a "Dionysiac abandon" (AD 102)—a sense of ecstatic surrender to currents of energy coursing through their bodies that propel them to move. Moreover, the undulating line conveys a sense that these currents are themselves never-ending, eternally impelling new movements in whatever media are capable of responding. As such, Duncan interprets, these undulating lines communicate *participation* in a "universal Dionysiac movement." As she explains: "One of the commonest figures in the Bacchic dances is that with the head turned backward. In this movement one senses immediately the Bacchic frenzy possessing the entire body. The motive underlying this gesture is in all nature. The animals, in Bacchic movement, turn back the head—It is the universal Dionysiac movement. The waves of the ocean form this line under a storm, the trees in a tempest" (AD 91).

As significant for Duncan in discerning this undulating line was what she did not see: the positions of the body idealized in classical ballet. She did not see legs extended above the ears, knees locked into a turned-out position, or arms fixed in one of five shapes. The Greek images seemed to express *movement*. They conveyed a sense of dance itself as a *generative* flow in which each movement empties itself into another without pause: "Every movement, even in repose, contains the quality of fecundity, possesses the power to give birth to another movement" (AD 90). These undulating lines, then, not only offer a visual, visceral experience of a "rhythmic unity which runs through all the manifestations of Nature" (AD 102), they do so in a way that presents the dancer who makes those movements as someone capable of sensing, responding to, and recreating images of such a rhythmic unity. The undulating lines, Duncan gathered present a portrait of the human bodily being as "one with the great movement that runs through the universe" (AD 68), and as the moment in that "one" where "it" may be grasped as such.

In short, what Duncan discerned in the undulating lines of Greek dancing echoes what she also found in Nietzsche's writings: a celebration of the inherent creativity of bodily beings. What she claimed to have seen in the Greek figures was not the truth of the body or of the infinite per se. What she saw were images—kinetic images—images that employ bodily movement as a medium in which humans generate images of themselves as participants in what their movements represent as a "universal Dionysian movement." In this way, the Greek figures expressed the Greek knowledge of a body's power to create and become images of highest value, beauty and holiness. In short, she found a vision for dance as "religion," as theopraxis, as an "expression of life." As Duncan confirmed time and again, this action of a body creating beyond itself was the "lost art" to whose revival she dedicated her life. She was not interested in copying the steps of the Greek figures, but in putting herself "in touch with the feelings that their gestures symbolized" (AD 139)—the feelings of ecstatic participation in a "divine continuity." She intended her dancing to communicate participation in this experience—to effect in dancers and spectators alike a "magic transformation" from which they emerged with the ability to embrace the most harrowing sorrow as a reason to love bodily life.

In pursuit of this vision, Duncan sought to discover new movements, "primary or fundamental movements" that "have within them the seeds from which will evolve all other movements, each in turn to give birth to others in unending sequence of still higher and greater expression, thoughts and ideas" (AD 56). She distilled four principles to help her in this process, each of which resounds with Nietzschean insight. Such movements must find their sources in nature, harmonize with the forms of the bodies making them, flow from an awakened soul, and fourth, function in the present day as the elemental rhythms of the chorus in Attic tragedy did for the ancient Greeks. As chapter 5 elaborates, these criteria guided her in developing an original approach to dance education, dance making, and performance that has philosophical import as a case study in revaluing Christian values, especially those aimed at women and their bodies.

Studying Nature

> *All true dance movements possible to the human body exist primarily in Nature.*
>
> —AD 69

In the end what Duncan claimed to have learned from her study of Greek art was not how to dance at all but how to study the natural world as a source for making dances that would convey an experience of participating in Dionysian rhythms of life. For Duncan, the fact that the Greeks were able to imagine the human body as incarnating a divine continuity attested to the attention they must have paid to the natural world "wherein all is the expression of unending, ever-increasing evolution, wherein are no ends and no stops" (AD 57). To Duncan's eyes the undulating movements express a relationship of a human being to the natural world characterized by observation and appreciation; by the Greeks' desire and ability to recreate what appeared to them in nature as beautiful in forms of their own making. As Duncan confirmed when landing upon Greek soil early in 1903, the ancient Greeks of "two thousand years ago" had "perfect sympathy and comprehension of the beautiful in Nature," a "knowledge and sympathy [that] were perfectly expressed in their own forms and movement" (IS 36). The Greek figures thus helped her *see* "nature" as a primary source of dance inspiration. She concludes: "in my art I have not at all copied, as is believed, figures from Greek vases, friezes, or paintings. I have learned from them how to study Nature" (AD 102).

In her own study of nature, Duncan not only observed natural phenomena, she read books about science, evolution, and electricity. Gathering her insights, she distilled the significance of the undulating lines she observed in Greek art into a theory of *wave movements*. By recreating the forms of waves, Duncan asserted, humans enact their creative participation in a field of energy common to all phenomena. She writes: "[A]ll energy expresses itself through this wave movement, for does not sound travel in waves, and light also, and when we come to the movements of organic nature it would seem that all free natural movements conform to the law of wave movement. The flight of birds, for instance, or the bounding of animals" (IS 45). That all energy expresses itself through wave movement, Duncan continues, attests to a common force acting on all phenomena—gravity. As a result, when a dancer discovers his wave movements in his own form he is discovering at the same time his connection to what is. When he discovers his ability to recreate and repeat the forms of waves as they appear to him, he discovers his humanness: his unique bodily-based capacity to experience, represent, and through this process *know* his participation in the constitutive forces of life. As Duncan writes: "All movement on earth is governed by the law of gravitation . . . by attraction and repulsion, resistance and yielding; it is that which makes up the rhythm of the dance" (AD 90). The implication is that a dancer who makes wave movements not only participates in a rhythmic force field that unites all of nature, he allows those who witness his dance to see and sense what is *as* a universal Dionysian movement, a divine continuity. He brings that continuity into being as real for himself and for others. And in the process, he becomes who he is: a moment in nature where "nature" reflects on itself and becomes what it is. When Duncan argues, then, that the movements of a dance *should* "follow the rhythm of the waves: the rhythm that rises, penetrates, holding in itself the impulse and the after-movement; call and response, bound endlessly in one cadence" (AD 99), the force of her "should" lies in what a dancer can feel and know and communicate: ecstatic participation in rhythms of a universal Dionysian movement creating through him.

The significance of this principle for Duncan as a guide in overcoming her physiological contradiction and revaluing religious values emerges in her application of it to the case of women. The study of nature, she insists, includes the study of the human body and its rhythms, form, and laws: "The human being too is a source" (AD 78). In so far as every body is a potential font of movement ideas, the fullness of human knowledge requires all bodies, female and male, to participate in discovering these movement potentials. As Duncan insists: "Woman is not a thing apart and separate from all other life organic and inorganic. She is but a link in the chain, and her movement must be one with the great movement which runs through the universe; and therefore the fountain-head for the art of the dance will be the study of the movements of Nature" (AD 68). In making such a claim, Duncan did not forget that most of the dancers in her time were women. She acknowledged that even in ballet, where the stars were women, the choreographers, impresarios, patrons, and beneficiaries were primarily men. Yet the movements the women performed on and off stage were for the most part not movements they themselves discovered through the careful study of and attention to their own bodily natures. For women to dance as an expression of life would be radical: they would renew religion by honoring their own bodily movement as the source of religious ideals, as having value.[9]

Duncan's call (for women) to study the movements of nature as the "fountain-head for the art of dance" has implications that extend beyond the discovery of seed movements per se. It implies that a person's sense of her own "nature," including her gender, is a function of her physical engagement with herself in movement.[10] It implies that by dancing, a person can generate kinetic images of (her) nature and affirm her particular bodily becoming as the locus for creating such kinetic images. It is in this sense, for Duncan, that dancing transforms a person's experience of herself: transforms how she moves and the value she accords that movement as constituting her relation to herself, others, and the world. As Duncan insists, "Dancing expresses in a different language, different from nature, the beauty of the body; and the body grows more beautiful with dancing" (AD 78). Such a dancer, in realizing the capacity of a body to become beautiful (that is, to incarnate ideals of what beauty is) through the process of making her own movements, is engaged, as Nietzsche envisioned, in theopraxis—a practice in which we create the gods who dance (through us).

Harmonizing Movement and Form

> The great and only principle on which I feel myself justified in leaning, is a constant, absolute and universal unity between form and movement; a rhythmic unity which runs through all the manifestations of Nature. The waters, the winds, the plants, living creatures, the particles of matter itself obey this controlling rhythm of which the characteristic line is the wave. In nothing does Nature suggest jumps or breaks; there is between all the conditions of life a continuity or flow which the dancer must respect in his art, or else become a mannequin.
>
> —AD 102

From her first principle—studying nature—Duncan distills a second that elucidates how a dancer who makes (kinetic images of) wave movements not only incarnates new values of bodily beauty and holiness, but also communicates participation in (her danced image of) Dionysian continuity. This second principle concerns the harmony of movement and form. What characterizes a wave, according to Duncan, and the secret of its undulating line, lies in the fact that the movement and the form are one. A wave *is* a form that is movement and a movement that is form. There is no wave without movement; yet that movement cannot appear without the medium in and through which it flows. As such, a wave is both moving and at rest. As Duncan insists: "It is this quality of repose in movement that gives to movements their eternal element" (AD 90). A dancer who makes movements that harmonize with his form, then, conveys a sense that his bodily form is a medium in and through which movement flows. A body whose movements harmonize with its form conveys a sense of continuity extending eternally through time and space.

The implications of this principle are several. First, wave-form movements allow us to experience the human body making those movements as *beautiful* (AD 71). When the movement of a body is one with the form of a body, then none of the body's movements appear awkward or forced. "It" appears as graceful, supple, strong. Second in so far as the movements of dance are those that harmonize with the particular form of the dancer, then the process of discovering such movements necessarily entails that a person direct her attention to her own form and develop a sense of it as moving. There can be no general formula for learning how to dance. Third, the process of learning how to dance will thus require that a person cultivate a new relationship with herself. In discovering and repeating movements that harmonize with her form, she will inevitably get better at making those movements. As her facility and agility improve, she will experience, concretely, her own ability to create and become images of herself, kinetic images. Moreover those images will represent—regardless of their particular form—the value she attributes to her body as she attends carefully to its form. Fourth, her dancing will thus enact this process—that is, both realize and represent this process—as something human, as what a human can do. Finally, in making movement that harmonizes with her bodily form, a dancer will project a vision of dance as that activity in which a person creates and becomes her highest ideals, participating in the universal, rhythmic continuity that is the source of all life. Such dancing will thus provide dancers and spectators alike with a visual, visceral knowledge of themselves as embodying what Nietzsche and Duncan identify as a Dionysian paradox: the movements of the dance will express the creative power of the individual dancing body *by* enacting its ecstatic dissolution in rhythms of "nature." A dancer appears as lost to and empowered by what her wave movements make visible and visceral in the form of an undulating line.

Awakening Soul

[T]he soul can be awakened, can completely possess the body.

—*AD* 52

A third principle, however, is crucial for understanding how Duncan moves from her theoaesthetics of wave movement to the creation of dance forms capable of renewing religion. As chapter 5 explores in greater detail, Duncan does not codify an array of wave form movements and then teach her students to make them. Such an approach to dance training would reinforce the dualistic sense of mind over body she intended her dancing to overcome. Rather, she seeks to awaken in her students a kinetic responsiveness, a *physical consciousness* of their own bodily selves as capable of sensing and recreating movement impulses. It is this physical consciousness that will provide her students with what they need to discover for themselves how to make movements that harmonize with their individual bodies. She describes this physical consciousness as "soul": "Awakening soul" is the "first step in dancing" (AD 52).

While this soul language initially appears to reify the dualism Duncan purports to reject, a closer look suggests otherwise. Duncan uses soul language as Nietzsche does—in ways that revalue the Christian conception of "soul" as a substance distinct from and of higher value than the "body." What Zarathustra teaches about soul is what Duncan believes: "soul" is first and foremost a word; it is a metaphor humans use to say something about the body (chapter 2). As such, it represents an image that a body creates of itself, of its essence or most valued self. By claiming that dancing is movement flowing from an awakened soul, in other words, Duncan reveals herself as one of Zarathustra's dancers: she can and does perceive soul as an expression of a physiological state. Dancing, she implies, in so far as it effects a transformation of bodily experience, is capable of giving birth to new meanings of "soul." In using soul language to describe dance, then, Duncan positions dancing as a practice for helping people overcome the "Puritan" reduction of the senses at work in western culture.[11]

In one of the most frequently recounted moments of dance history, Duncan describes the experience in which she claims her "soul" awoke and she discovered the source of the religion-renewing dance she desired. Sometime in the course of her pre-Nietzsche European travels (1900–1901), Duncan was standing in her studio listening, hands across her chest, her mother on the piano. She was waiting to feel an impulse to move in response to the music. Even to undertake this exercise expressed a Nietzschean faith in the body: she believed she would sense something.

She did. What happened (or so she reports) was that an impulse to move arose in her solar plexus and radiated out through her limbs, propelling her through space. Her "soul" awakened. She experienced this "soul" in the same moment that she sensed her solar plexus as its "temporal home" (ML 341). Looking back on this moment, Duncan believed she had discovered the source of movements that would both harmonize with her form and enact her participation in the constitutive rhythms of life. As she writes, the experience crystallized the "first basic theory of my Art" (ML 76): dance is movement flowing from an awakened soul.[12]

In Duncan's verbal elaborations of this moment, it is easy to lose the radical edge of her claims. For example, she speaks of seeking "the divine expression of the human spirit through the medium of the body's movement" (ML 75), suggesting to some readers that she believes in a "spirit" (or soul) existing apart from the body, and then looks for gestural signs to represent it. However, the medium Duncan identifies in this quotation is not "the body" but a body's *movement*, and the human spirit does not appear as "divine" until it finds expression in that movement. Said otherwise, the fact that Duncan feels an impulse to move and feels it in the tangle of beating, breathing rhythms that comprise the solar plexus proves to her what she believed to be true from her study of Greek art: it is bodily movement that provides humans with a medium for gathering and expressing knowledge of their relationship to what appears to them in and through that movement as a universal Dionysiac movement. To say that "soul" awakens, then, is to say that a person develops a physical consciousness of her kinetic creativity.

This principle of soul-awakening further illuminates the connection discussed earlier that Duncan makes between the study of undulating lines in nature and art and the rebirth of Christian values. When Duncan studies wave movements, she does so through her own awakened physical consciousness. Her study is not primarily an intellectual exercise. She does not identify forms, conceptualize them into patterns, and then coax her body into making those shapes. Rather, she animates what Nietzsche described as the primary level of metaphor-making. She orients herself in relation to a particular phenomenon or environment, opening and attuning her awareness. She actively resists the "Puritan" trend of reducing the senses to nil. She listens, but not with her ears per se. She listens in and through the space of physical consciousness she has discovered and cultivated within herself. In so far as she senses any impulses to move welling within the field of her sensible awareness, she confirms her belief that she is capable of such response. As she learns to follow these sensed impulses through her body, the form of her body, its potential and reach, also appear to her with greater clarity. She discovers what her body can do. In this way, then, Duncan's claim that dance is movement flowing from an awakened soul represents a radical indictment of the ascetic ideal. She is claiming for bodily movement a constructive role in the creation and realization of our highest ideals, our selves. She is framing the practices of studying nature, recreating wave forms, and finding movements that harmonize with one's particular body as practices through which humans realize the potential of their bodily creativity and become fully human. In doing so, she is framing dancing—her vision for dance—as an art that not only generates new values but itself embodies an alternative morality, an alternative physical–spiritual discipline, to that represented by the Christian teachings she rejects. Dance rendered as movement flowing from an awakened soul is akin to the dance envisioned by Nietzsche as theopraxis.

Dionysus Reborn

> To unite the arts around the Chorus, to give back to the dance its place as the Chorus, that is the ideal. When I have danced I have tried always to be the Chorus: . . . I have never once danced a solo. The dance, again joined with poetry and with music, must become once more the tragic chorus. That is its only and its true end. That is the only way for it to become again an art.

—AD 96

Finally, evidence that Duncan perceived her project as participating in Nietzsche's revaluation of values emerges in a fourth principle she developed to guide her discovery of religion-renewing, woman-liberating seed movements: *be the chorus*.[13] In several essays where Duncan lifts Nietzsche's account of Attic tragedy nearly intact, she uses "soul" language to make the connection between her vision for dance and the role played by the chorus: she likens her experience of soul awakening to Nietzsche's account of the "magic transformation" effected by choral dancing and singing. Duncan thereby identifies her vision for dance (flowing from an awakened soul) as the key to realizing the alternative *faith*—the radical affirmation or love of life—that Nietzsche calls upon "Dionysus" to represent.

Duncan makes the connection between soul-awakening and magic transformation when she describes the "Chorus" as "the very soul of tragedy" (AD 84). In Duncan's reading (which informed chapter 1), the appearance of the chorus in the course of the dramatic narrative marks the moment where a tragedy delivers to spectators a visceral, visual sense of losing their individuality to the constitutive rhythms of the universe. Spectators emerge transformed. As Duncan confirms, speaking of Aeschylus: "Entering at the most poignant moment of emotional tension, it [the chorus] brought a lyric exaltation, the eternal and divine point of view.

Deepest soul of Tragedy: the Chorus was Wisdom or Reason or Joy or Sorrow eternal" (AD 92). In this ecstatic moment of "lyric exaltation," the "elemental rhythms" of dancing and singing compel mimetic response, driving a spectator to identify with (an image of himself as) a god—"the eternal and divine point of view." Spectators feel free, immortal—they overflow with a vibrant sense of their own health. And this feeling of strength provides them with a perspective—a transformed physical consciousness—from which to *affirm* the fate of the tragic hero as a *necessary* moment in the rhythmic pulse of life. Duncan summarizes: the "soul of audience, harrowed to the point of agony, was restored to harmony by the elemental rhythms of song and movement. The Chorus gave to the audience the fortitude to support those moments that otherwise would have been too terrible for human endurance" (AD 84).

At the same time, by designating the chorus as the soul of tragedy, Duncan acknowledges that the chorus is effective not only by allowing spectators a physical consciousness of themselves as an eternal and divine point of view, but by providing them with an experience of their own participation in the creation of the experience of this divine point of view. As Duncan discerns, the chorus is necessarily comprised of individual dancers, moving their bodies in patterns that are planned, rehearsed, and in unison. The ability of the chorus to deliver what appear to be "elemental rhythms" presumes the ability of individual humans to discover and recreate movement patterns that harmonize with their human forms. The chorus' bodily movements, in other words, communicate a "double message of Apollo and Dionysus" (AD 140) as described in chapter 1. Choral members are effective in conducting an experience of (Dionysian) dissolution into a divine point of view, but only in so far as they exercise and confirm the (Apollinian) ability of humans to do just that—that is, make and incarnate kinetic images of their relationship to (what appears in their kinetic images as) divine continuity.

When Duncan claims to "be the Chorus," then, she is not suggesting that she is more than one person, or even that she is anonymous. To be the chorus is to serve as a catalyst for communicating the "double message of Apollo and Dionysus": to catalyze in spectators a physical consciousness of themselves as creative participants in the constitutive rhythms of the universe. To be the chorus is to dance from an awakened "soul" (as Duncan revalues it); and it is to dance in ways that awaken the souls of those willing to watch, witness, and respond. It is to awaken in dancers and spectators the physical consciousness that provides them with the resources they need to affirm life even in the face of suffering.[14] As such the chorus catalyzes an experience in which a person acknowledges and overcomes a physiological contradiction—the fact of being both an individual body and lost to nature—by embracing that tension as an occasion to sing, dance, and participate bodily in the creation and realization of highest ideals. As Duncan confirms, the chorus is "the eternal hymn of the struggle between man and Destiny" (AD 82). In being the hymn of this struggle, the choral singing and dancing encapsulates the rhythmic becoming of *life*. As Duncan writes approvingly of Nietzsche's dance: "he did not mean the execution of pirouettes. He meant the exaltation of life in movement" (AD 77).

Finally, in insisting that she intends to be the chorus, Duncan also concedes the limits of what dancing alone can accomplish. Choral dancing works to transform physical consciousness only when a person has been made aware, by the harrowing tale of the visual and narrative imagery, that such a transformation is needed. Dance needs the other arts in order to accomplish its mediating role. As Duncan asserts, dance fails if it "aspires to be everything, to take the place of poetry and drama"; in trying to imitate the spoken language, dance degenerates into "pantomime" (AD 95). It fails to communicate the Dionysian dimension of its double message. For this reason, then, dance must cooperate as one among the other arts to succeed in renewing religious attitudes toward embodiment. "[O]ur theatres will become temples" Duncan writes, "when music, poetry, and architecture all unite around the dance, highlighting its distinctive contribution to the project of value creation" (AD 135). In dancing by herself to classical music as she did, Duncan was taking up one part of the vision she shared with Nietzsche, seeking to discover seed movements whose undulating lines would stimulate the work of other artists to complete the picture of a renewed religion.

A Conception of Life

> For me the dance is not only the art that gives expression to the human soul through movement, but also the foundation of a complete conception of life, more free, more harmonious, more natural. It is not, as is too generally believed, a composition of steps, arbitrary and growing out of mechanical combinations—which even if they serve well as technical exercises cannot pretend that they constitute an art. This is the means, not the end.
>
> —AD 101

The significance of these four principles thus described extend far beyond a method for creating dances. The stakes are moral, existential. Duncan is offering a philosophy of religion and dance. While it may be said that every dance expresses some conception of life, dancing that aligns with Duncan's principles presents dance itself as the locus of our god-forming instinct—as theopraxis. It does so not by yoking dancing to a particular religious tradition; nor by paying allegiance to a religious institution, nor by treating the theme of supernatural entities. Rather, a dance that embodies her principles necessarily engages and revalues attitudes toward soul and body, human and God, female and male, predicated on the ascetic ideal. The undulating lines of such dancing not only express the inherent creativity of bodily becoming, they affirm the *practice* of dancing as a transforming experience that enables a person both to generate an *ideal* of "universal Dionysiac movement," of Nature as divine, *and* to know that ideal as true for him. Duncan intended dance to develop in us the physical consciousness and vitality we need to greet whatever occurs as an opportunity for exercising the creative power of our bodily becoming. In short, dancing that incarnates Duncan's principles expresses a conception of life in which dancing itself is valued as a medium in and through which we become who we are.

In this reading, whether a "Nature" or "divine continuity" actually exist apart from her dancing is not Duncan's concern. Her point is that dancing that follows her criteria—in making movements that harmonize with one's individual form—enables certain kinds of knowledge that are not otherwise available. This knowledge that dance provides, moreover, is precisely what modern humans need to embrace and overcome the physiological contradictions they live between their indebtedness to ascetic ideals and their resistance to them. It is knowledge that affirms bodily differences—whether coded by gender, race, class, and so on—as significant sources of human value that should be honored as such. It is via such dancing, Duncan insists, that we come to know what it is that sanctifies us: the movement of our own bodily becoming.

Dancing into the Future

To revive the antique dances would be a task as impossible as it would be useless. The dance, to be an art for us, must be born out of ourselves, out of the emotions and the life of our times, just as the old dances were born of the life and the emotions of the ancient Greeks.

—AD 139

In the context of the vision she shares with Nietzsche, Duncan's call for a great renaissance of religion, her claim that dance is a religion, her use of terms like "soul" and "Dionysus," all appear in a new light. What I am calling Duncan's *religion language* represents her participation in Nietzsche's revaluation of values. What has appeared to many commentators as confused and contradictory statements about religion here appear as evidence that in her dancing, Christian attitudes toward bodily being are being overcome and new values are emerging.

Seen from this perspective, Duncan provides us with a concrete example of what Nietzsche's project of revaluation entails. Duncan does not deny the fact that she lives a physiological contradiction between her (Christian pagan-inspired) gods of beauty and love and her (pagan Christian-inspired) celebration of bodily becoming. Rather, she animates the tension inherent in this contradiction to generate forms of dancing that redefine the value of both religion and dance. Her vision for dance redefines "religion" as an expression of bodily becoming whose forms must be held accountable to the physical consciousness and health it enables. Her vision for dance redefines "dance" as a therapeutic theopraxis—an activity in and through which we create and become our highest ideals. In these ways, then, Duncan overcomes the physiological contradiction she lives. She does so by changing the value it has for her: she embraces it as the condition that has enabled her to develop the empathetic insights and critical perspective she needs in order to *be*, in her *dancing*, the evolution of Christian values along the trajectory of commitment to what she shares with them, a love of life. She can envision life-affirming values because she knows in her own lived experience how life-negating values function, how they train our senses in ways that encourage ignorance of the body and of dance, and why they do not deliver the salvation they promise.

Duncan was not content, however, simply to envision what a dance of the future might look like. She created a system of dance education and choreographed a body of dances that function as the kind of extra-textual work to which Nietzsche's dance images point. It is in her practice and performance of dance that Duncan incarnates new values and embodies new philosophical perspectives concerning the relationship of religion and dance.

Chapter 5

Incarnating Faith

People have never understood my true aim. They have thought that I wished to form a troupe of dancers to perform in the theatre. Certainly nothing was farther from my thoughts. Far from wishing to develop theatre dancers, I have only hoped to train in my school numbers of children who through dance, music, poetry and song would express the feelings of the people, with grace and beauty.

—AD 117

It is difficult to say which criticism of Duncan has been more common—that she failed to develop any dance technique, relying instead on her charismatic powers, or that her career in dance was undermined by her lack of moral discipline.[1] Such responses did not surprise her. From the perspective of those whose values she aimed to renew, her work would, of course, appear as neither moral nor dance-like.[2] Such responses confirmed what Duncan believed even in their judgment of her: that whether and how we dance is a *moral* issue. Whether and how we dance expresses the value we accord to our bodies, our fellow humans, and the farthest horizons we can imagine. As Duncan insists, to criticize her for failing to abide by the mores of Christian society or the canons of classical ballet is to hold her accountable to goals and values she rejected. It is to miss what her devotion to Nietzsche's writing illuminates: her triple vision, her commitment to renewing religion through dance in ways that liberate women.

Evidence that this commitment lay at the heart of Duncan's dance process appears not only in the principles of her dance philosophy, as described in chapter 4, but also in the concrete actions she undertook to realize her visions. Even though Duncan toured and performed as a professional dancer, she perceived such touring and performing as a means; her goal was to found a government-sponsored *School of Life* and not dance where all children, including those who could not otherwise afford it, could learn to *live*, to *express life*, by practicing dance. Every stage of her dance process—her practice, choreography, and performance—evolved to meet the needs of students in her school. That her successes in touring paved the way for the advent of a highly professional art form, then, cannot be held against her. Rather, in order to understand the power and potential of the modern dance she helped inspire we need to remember what she accused her critics and imitators of forgetting: that *dance* is movement flowing from an awakened *soul*. To dance, as Duncan envisions it, *is* to participate in the revaluation of Christian values toward (female) bodies.

A School for Life

To rediscover the beautiful, rhythmical motions of the human body, to call back to life again that ideal movement which should be in harmony with the highest physical type, and to awaken once more an art which has slept for two thousand years—these are the serious aims of the school.

—*AD 132*

Duncan's commitment to participating in such a revaluation is evident in her repeated efforts to build her school. Only two years after her first commercial success, in 1904 (until 1908), she welcomed twenty-five girls into its first incarnation in Grunewald near Berlin. Later incarnations appeared in Darmstadt, Germany (1911, opened by her sister Elizabeth); on the outskirts of Paris in Bellevue, France (1914, named *Dionysion*), and in Russia (1921–1922). Receipts she earned from touring she plowed back into the life of the school; when finances ran low, she accelerated her performance schedule.

As suggested, Duncan's aim in founding a school was not to groom individuals for the stage. When some students of hers left to go on "recital tours," she was critical of them: they were accommodating their art to the terms offered by western culture. Duncan firmly believed that dance could not realize its potential as art if dancers abided by the false dichotomy between dance and religion institutionalized in the chasm between stage and church. As she writes, her students had "forgot[ten] their mission and left the group, to follow impresarios who were ready to exploit them and to take them on recital tours through all the world" (AD 117). Their "mission," as Duncan saw it, was to realize the potency of dance as art. Yet, within the terms and venues set by the impresarios, audiences would be led to perceive their dancing as entertainment and diversion. Their dancing would appear as one more commodity in the marketplace, and not the *expression of life* Duncan intended. Duncan herself refused invitations to perform in traditional dance venues, preferring concert halls where orchestras performed classical music and where audiences expected to witness edifying art. Instead of professional dancers, then, Duncan wanted her students to become teachers. She wanted to train dancers who would reject the terms within which dance was being made, performed, and received, and teach others to do the same. They would thereby catalyze a renewal of Christian values toward dance and embodiment.

Duncan's commitment to revaluing values is also evident in her choice of whom (and not just why) she taught. Duncan insisted on teaching children not adults. Children, she confirms, have not yet succumbed to the forces of desensualization and oversensitization characteristic of modern western culture. As such, they have not learned to ignore their sense of kinetic responsiveness; they retain a capacity that most adults have lost for appreciating dance as an expression of life. Duncan explains: "In childhood we feel the religious sense of movement poignantly, for the mind is not yet clouded with dogmas or creeds. Children give themselves up entirely to the celebration and worship of the unknown God, 'Whatever gods may be.' In fact a child can understand many things through movement of its body which would be impossible for it to comprehend by the medium of the written or spoken word" (AD 124). Because children are not yet bent to practices of writing and speaking, they have not developed the power of their minds to enforce dogmas or creeds over and against their bodies. They are still able to feel "an inner self awakening deep within" as the "strength" that enables them to lift their heads, raise their arms, and walk "slowly to the light" (AD 52). It is those without formal training in religion, Duncan insists, who understand the "language of the soul" (AD 52). Thus, her choice to teach children marks her attempt to intervene strategically in systems of education, religious and otherwise, that encourage people to perceive their bodies as material objects that should be controlled and directed by a more powerful mind.

That she deemed such intervention necessary confirms an insight she shared with Nietzsche about the nature and process of revaluing values. Christian values are us. It is not possible to refute them with even the best of will power. Likewise, it is impossible to generate new ideals until and unless we relearn our relationship to the "small things," the enabling details of our physiological health. Correlatively, while Duncan chose to work with children because they could respond kinetically, she did not aim to preserve the forms of their movement. She sought to help her students grow and refine their kinetic sensibility into a health-enabling physical consciousness, and one centered in the solar plexus. As Duncan affirms, "when I have taken children into my schools I have aimed above all else to bring them into a consciousness of this power within themselves, of their relationship to the universal rhythm, to evoke from them the ecstasy, the beauty of this realization" (AD 52). With physical consciousness of this "power within themselves," Duncan perceived, comes an ability and responsibility to exercise it. She wanted to help children develop the capacity to participate in the bodily process by which humans discover, recreate, and become kinetic images of their relationship to the "universal rhythm." To catalyze such bodily becoming was Duncan's intent.

Given this program, the training Duncan offered in her school would not have appealed to dancers striving to join one of the ballet companies in Europe. Students did not learn technical wizardry; they did not experience the rigor of conforming their bodies to the codes of a classical form. What Duncan offered would have appeared nearly invisible to those seeking such exercises. She sought to create the conditions—intellectual, physical, emotional, and spiritual—for helping children cultivate physical consciousness, discover movements that harmonize with their individual forms, and thus learn to value their bodies as inherently creative agents of human becoming, as "beautiful" and "holy." In her awareness of how important dance education is to the ideas we think, the values we hold, and the experiences we open to receive, her philosophy aligned with Nietzsche's.

Practice

The practical means Duncan developed for helping children discover the "power within themselves" flesh out the criteria Nietzsche associates with dance in his later works (see chapter 3). Duncan designed exercises to help her pupils strengthen their instincts, open and refine their senses, and generate and invest their energy in creating and realizing ideals in the form of kinetic images. Duncan's school was a place where children lived in a world whose defining characteristics she selected as the most conducive to awakening (her revalued understanding of) *soul*. I describe three of these elements.

Gymnastics Exercises

That Duncan incorporated gymnastics exercises as part of her school may seem contradictory at first, given that Duncan railed against Swedish and

German "gymnastics" for focusing exclusively on the "development of the muscles." While she admitted that gymnastics offered some advantages over ballet training (in that it did not involve distorting the body to the same degree), she was still concerned that the practice of gymnastics reinforced an experience of self as a mind acting to make the body into what "I" want it to be—namely, strong. She writes: "German and Swedish gymnastics have in view only the development of the muscles; they neglect the proper coordination of spirit and body . . . a force from the outside directs the movements under control of the will" (AD 119). In so far as practicing gymnastics exercises the will as a force over and against the body, it impedes the "proper coordination," or interdependent responsiveness, of what people come to know in and through their actions as "body" and "spirit" or "soul." In other words, for Duncan, by practicing such exercises people create themselves into persons who perceive themselves and the world in dualistic terms.

Nevertheless, even through Duncan rejects the "development of muscles" as a goal, she does not reject the development of muscles as a *means*. Physical consciousness itself requires an ample degree of bodily vitality and health, strong instincts and senses. As she avers, "it is necessary to give the body plenty of air and light; it is essential to direct its development methodically. It is necessary to draw out all the vital forces of the body towards its fullest development. That is the duty of the professor of gymnastics. After that comes the dance. Into the body, harmoniously developed and carried to its highest degree of energy, enters the spirit of the dance" (ML 174–5). "Methodical" direction of development is essential for "harmonious" development of a body because a body is an ongoing becoming. Bodily becoming does not happen in a vacuum; a body cannot *not* be influenced by the contexts in which it grows. Thus Duncan used gymnastics exercises to challenge students physically, to enliven their sensory perception, and to spur them to find and tap sources of creative energy within themselves.

Nevertheless, the language Duncan uses to defend the practice of gymnastics as integral to the process of dance training still seems to recapitulate the dualistic conception of self she purports to reject. She asserts, for example, that the movements of a gymnast express the *body*, while the movements of a dancer express "the sentiments and thoughts of the *soul*" (ML 175). From a Nietzsche-enabled perspective however, another interpretation appears. Duncan's claims serve to expose the Christian morality embedded in the practice of gymnastics, and usher in dance as the vehicle of its revaluation. The movements of the gymnast express "the body" in the sense that they express a belief that the body is a material object in need of externally imposed disciplines. By contrast, the movements of the dancer express a belief that the body is a movement of its own becoming, generating images of itself, including the sentiments and thoughts we associate with "soul." In making this contrast, then, Duncan is framing dance as a bodily activity in and through which "soul" comes to life. She is thus again revaluing "soul" as a something a body creates for itself in and through its own movement practices. A dancer, for Duncan, is one who perceives her bodily materiality neither as an obstacle to spiritual growth, nor as an ignorant mass requiring control by an agile mind, but as a dynamic locus of (self) revelation. Thinking about and practicing *dance* as an expression of *soul* in this sense serves to dislocate and dissolve dualistic perceptions of body and mind, or human and God.

A second example of how Duncan evokes soul language in an effort to revalue Christian values appears in her descriptions of how gymnastics exercises render the body an "instrument": "The nature of these daily exercises is to make of the body, in each state of its development, an instrument as perfect as possible, an instrument for the expression of that harmony which, evolving and changing through all things, is ready to flow into the being prepared for it" (ML 175). Here the "instrument" Duncan has in mind is not an object played by a mind or soul that exists apart from the body. Body is an "instrument" in relation to currents of energy coursing through the universe. As an "instrument" a body is capable of sensing and amplifying—of being played by—the smallest movement impulses arising within its field of physical consciousness. As an instrument, in other words, a body embodies its own intelligence; it knows how to translate energy into art; it is the process of this transformation. To make such an instrument of a body, moreover, persons must learn to empty their minds into the spaces of their physical consciousness, and develop the sensory properties unique to their bodily being. Thus, to claim that bodily movement is an instrument for expressing "sentiments and thoughts of the soul," is to claim that it is through the *practice* of dancing that people acquire and exercise an ability to create beyond themselves. In the practice of dance a body becomes what the act of exercising its own wisdom allows it to be. It is "daily exercises" that make of the body an "instrument" through which persons overcome the lingering effects of Christian morality on their lived experience, and affirm their bodily becoming as a locus of revelation.

The "gymnastics exercises" to which Duncan introduced her students were of the most basic kind: walking and running, skipping and leaping. As Duncan explains, "The pupils . . . run and jump naturally until they have learned to express themselves by movement as easily as others can express themselves by word or song" (AD 82). Again the "naturally" here is easily misunderstood. Duncan's point is that students must *learn* to move *naturally*. They must learn to make movements that are fundamental to any human life. A Duncan teacher provides clear instructions for how to walk, skip, or run in ways that allow students' movements to align with the mechanics of their bodies, increase the resilience of their ankles, maintain alignment of their knees, and open up the capacity of their hips. They learn to adjust their own movements from within based on how they feel when executing them. Students thereby learn to do movements they may already do but with an awareness of how to make those movements in ways that nourish rather than impede a physical consciousness of their own bodily becoming. Thus, in so far as they learn to "express themselves by movement," pupils are not learning how to correlate mental states with a gesture vocabulary; they are learning how to access and exercise a coordination of physical and mental awareness. They are acquiring a physical consciousness of how bodily movement is a medium for "expressing"—in the sense of *becoming*—themselves.

Atmosphere of Beauty

> Many years ago the idea came to me that it might be possible to bring up young girls in such an atmosphere of beauty that, in setting continually before their eyes an ideal figure, their own bodies would grow to be the personification of this figure; and through continual emulation of it and by the perpetual practice of beautiful movements, they would become perfect in form and gesture.
>
> —AD 80

The revaluation of Christian values Duncan intends in her approach to teaching gymnastics finds support in a second aim of her school: to surround pupils with "an atmosphere of beauty." Often the children practiced gymnastics outside, on a lawn, surrounded by plants and trees, clouds, and breeze. Or they practiced to classical music in rooms decorated with images from the history of art, often images of dancing bodies.[3] By encircling her pupils with phenomena she found beautiful, Duncan sought to provide them with stimuli that would encourage them to identify viscerally, imitate gesturally, and respond kinetically to impulses arising in and as their physical consciousness. In such an atmosphere, Duncan reports, children develop a knowledge of themselves as responsive creators, a knowledge that they would not otherwise have. In learning to observe the "movements of the clouds in the wind, the waving of trees, the flight of birds, the whirling of leaves" they "develop a secret sympathy in their souls, unknown to others, which makes them comprehend these movements as most people cannot. For every fiber of their bodies, sensitive and alert, responds to the melody of Nature and sings with her" (AD 82).

As these comments suggest, Duncan did not encourage students to "imitate" nature; nor to copy the movements of clouds or trees. Rather students "observe" the qualities of movements they perceive around them, and are invited to sense movement responses in themselves. Encouraged to follow their movement impulses, they develop this sense of "secret sympathy": they sense their common participation in rhythms of energy that "Nature" also represents. As Duncan confirms: "How often, returning from these studies, coming to the dance room, have these pupils felt in their bodies an irresistible impulse to dance out one or another movement which they had just observed! A Dionysian emotion possesses them" (AD 82). Surrounded by an atmosphere of beauty, then, students are drawn into their senses. They grow more alert; more responsive and attuned to their responses, and more capable of experiencing the desire to move within themselves as a source of joy and self-knowledge, rather than as a source of sin, temptation, or terror.

Moving from the Solar Plexus

> Imagine then a dancer who, after long study, prayer and inspiration, has attained such a degree of understanding that his body is simply the luminous manifestation of his soul; whose body dances in accordance with a music heard inwardly, in an expression of something out of another, a profounder world. This is the truly creative dancer, natural but not imitative, speaking in movement out of himself and out of something greater than all selves.
>
> —AD 52

It is in the third trait of her school that Duncan's commitment to revaluing values appears most clearly. Students, physically vitalized by movement exercises, whose desire to move has been stirred by the inspiration of art and nature, are primed for the first step of dance as described in chapter 4: the awakening of soul. For Duncan, this awakening is critical: it marks the difference between dance as art and dance as entertainment, between dance as religion and "mere merchandise" (AD 62). It is a transformation from which people emerge with a physical consciousness strong enough that they can participate in making kinetic ideas of themselves that affirm rather than deny their bodily becoming as a source of knowledge and value.

The challenge, Duncan admits, is that such an awakening of physical consciousness cannot, strictly speaking, be taught. As she admits, dance is "to me an instinctive thing born with me" (AD 129); "The dances of no two persons should be alike" (AD 58). As such, Duncan designs movement exercises not to teach steps, but to help students open pathways of energy and consciousness in their bodies so as to enhance their ability to sense movement impulses arising within themselves. Specifically, Duncan's exercises train a dancer to notice impulses arising in the solar plexus, and to follow those impulses as they radiate outward through bodily limbs. Thus, even though Duncan professes that: "I have no system. My only purpose and my only effort have been to lead the child each day to grow and to move according to an inner impulse; that is, in accordance with Nature" (AD 119), her efforts to "lead the child" were systematic and comprehensive. To move in accord with "Nature" according to an "inner impulse" is not easy. Success is not guaranteed. A person must develop the discipline required to attend to her bodily movements and listen for her own kinetic responses in order to sustain an awakened physical consciousness and exercise the "power within."

The first movements in a Duncan class are indicative of her system-less approach. A dancer begins with arms stretched upwards toward the sky, fingers lightly unfurled. Chest is lifted and airy, feet and legs are rooted into the floor strongly and comfortably at hips' width. With the phrasing of the music, finger tips slowly begin to curl forward, triggering a cascade of energy down through the body. Hand, wrist, and elbow follow the fingers, reaching up to curl down. Arm, head, neck, chest, torso, waist, spill over in turn, doubling the dancer around the solar plexus until the dancer's hands rest gently on the floor. Knees bend and straighten, bend and straighten, until deep in the solar plexus the wave returns. The dancer slowly rolls up, unfolding into an upright posture as arms arc back up and overhead and fingers open to channel the rays of the sun.

As the wave movement repeats at least four times, a dancer clarifies a sense that his body is the wave it is making. With each repetition his ability to feel the wave moving through him, down and up, up and down, increases. His experience of his body changes; his body dissolves in a movement that moves itself. In making the movements he thus invites a physical consciousness of his self as a body who is what it is by virtue of (participating in) wave movement.

Another exercise further elucidates the revaluing of bodily values Duncan designs her practice to accomplish. After the roll downs, a dancer begins a series of exercises involving arm movements. One arm begins. With wrist loose and hand lightly dropped, the arm rises. The fingers trail a line along the central axis of the body, passing through the solar plexus and up in front of the throat, face, and forehead. The arm opens up over the head and arcs out to the side, where it floats down again to the side of the body. The left arm repeats the gesture, with forearm, wrist, and fingers passing up along the central cord and opening to the sky. In the next movement, as both arms repeat the movement together, the meaning of the exercise emerges. As the wrists cross and mount the lightly-held hands encircle the solar plexus, marking the opening of a passage, a site through which energy flows into and through an individual body. As both arms lift open and hands part, the energy crests and circles back around the lifted, opened chest. The arms trace a trajectory of movement that moves through the solar plexus as source and telos.

While useful for limbering the shoulder joints, these exercises mean more when interpreted in relation to Duncan's philosophy of dance and religion. The exercises not only help a dancer draw into consciousness a physical sense of solar plexus as a place where she might receive movement impulses (the "temporal home of the soul"), they provide her with resources for responding to such impulses. While the movements do not and cannot themselves awaken soul, repeating them increases the likelihood that if and when soul awakens, a person will notice, respond, and discover through that awakening how to move in ways that nourish the ongoing health and kinetic responsiveness of her bodily becoming. The fruits are evident. Practicing such attention to the solar plexus produces a suppleness in the joints, a soft, weighted connection to the ground, and a strong sense of movement flow through the body characteristic of Duncan dancers.

In sum, against her critics, Duncan is far from urging a laissez faire approach to making movement. She guides her students in the practice of a closely disciplined attention to the ways in which movements rise and flow through their bodies. " 'Natural' dancing," as she insists, "should mean only that the dance never goes against nature, not that anything is left to chance" (AD 79). To "never go against nature" presupposes that a dancer knows what nature is, and specifically, that she knows the "nature" of her own bodily becoming. Such knowledge is learned, or at least, not forgotten, and then cultivated. Moreover, such learning of dance does not always feel good. A newly discovered movement may feel awkward and uncoordinated. A body aches with the effort. Duncan does not deny these moments. The key for her is that the process of hurting and learning remain open and accountable to a sense of movement flowing through the body. The movement that results from such attention, she avers, will not be "left to chance"; it will express a human relationship to embodiment in which a person values her bodily movement as a medium for generating and becoming kinetic images of herself in relation to nature, for becoming human. While inspired by nature, "the dancer's movement will always be separate from any movement in nature" (AD 79).

When Duncan insists: "A child ought to dance as naturally as a plant grows" (AD 119), then, she is not forgetting that the ability of a plant to bear fruit depends on whether life-enabling conditions are present or not. In creating a School of Life where children engage in gymnastics and dance exercises surrounded in an atmosphere of beauty so as to develop their instinct for movement into a mature and creative physical consciousness, she was taking on this challenge: she would provide her students with the conditions they would need to be and express life-affirming values. She would create the contexts within which individuals might develop the energetic perspective needed to overcome and revalue ascetic ideals. While what she aimed to awaken in her pupils was, she admits, "incommunicable" (IS 47), her reasons for doing so were ones that her relationship to Nietzsche's project predicts. A dance capable of revaluing Christian values toward embodiment, in order to succeed, must express faith in bodily movement; it must express love for the body, and a willingness to wait and listen and trust in one's own body, one's own capacity for kinetic response, as a source of insight and value, of art and religion.

Dance Making

Even though Duncan was not interested in training professional dancers, she did want her students to be able to make and perform good dances. Making "godlike" movements under the influence of Dionysian emotions alone does not express faith in bodily movement as a medium for *communicating* knowledge. Said otherwise, for dance to renew religion more is required than a personal or even communal practice for developing physical consciousness. It is necessary to use the movements that one discovers in the practice of dancing as material for making dances that communicate participation in the experience of awakening soul.

Duncan identified at least three steps in her approach to dance making, and they parallel her approach to dance practice. First, she opens her senses to beauty in a manner she describes as reminiscent of all artists (AD 52). Second, she attends to the movement responses that arise in relation to the ideas, forms, themes, and images she perceives as compelling her to dance. She remembers, recreates, and repeats her movement responses in ways that allow the energy of her initial encounter to flow through her in different ways. She then uses these impulse–responses as the movement material with which she works to build the kinetic imagery of a dance. Describing these first two steps, she writes: "The true dancer, like every true artist . . . starts with one slow movement and mounts from that gradually, following the rising curve of his inspiration, up to those gestures that exteriorize his fullness of feeling, spreading ever wider the impulse that has swayed him, fixing it in another expression" (AD 99). Of note in this account is what Duncan's dance-maker follows. She follows the impulse–response that has swayed her—the "rising curve of . . . inspiration." She allows the gestures to lead her, from one to the next. She exercises the physical consciousness of undulating lines honed in her practice. In these ways her practice and her dance-making work together: the more pathways of energy a dancer has blazed through her physical consciousness, the more resources she has for "spreading ever wider the impulse" that has swayed her.

A dance is not complete, however, until a dance-maker performs a third task: he must shape the flows of kinetic imagery into the form of a dance. In this task, Duncan's watchword is "simplify." She writes: "In art the simplest works are those which have cost the most in the effort for synthesis, of observation and of creation; and all the greatest masters know what is the cost of true accord with the great and unrivaled model that is Nature" (AD 103). As this passage suggests, Duncan does aim to *make* dances that *look* natural, and the process is hard work. A dancer must discard many of his discoveries. He must discipline himself to his material and honor its logic. However the result is worth it. The dances that emerge from such a dance-making process are reflexive: they express the physical consciousness a dancer exercised in discovering movement, the kinetic imagination he used in recreating the forms of his inspiration, and the act of reducing his creations, disciplining himself to nature, which the final production of the dance required. Such dances, in other words, revalue the ascetic ideal. They represent an overcoming of the physiological contradiction of being both Christian and anti-Christian for in them the ascetic ideal has value as a guide to honoring and perfecting the god-making potency of *bodily* becoming. And this action of dance making, for Duncan, differs only in degree from what any person does in living every day.

Blue Danube

Two examples from Duncan's repertoire shed light on her process of dance making and its relation to Nietzsche's project of revaluing ascetic values. In *Blue Danube* (1902), one of Duncan's earliest and most popular dances, she seems to do what critics assail her for: imitate a river. As described in the Introduction to Part 2, Duncan's movements sweep along with a powerful grace, dipping and twirling, reaching, up and back, from left to right, evoking the formal rhythms of waves. In what sense could this dance enact an overcoming of physiological contradiction, or a renewal of religion?

When viewed through her (Nietzsche-inspired) principles of dance, answers appear. In this dance, Duncan is not imitating water; she is recreating the movement impulses stirred in her solar-plexus by her attentive study of water. The kinetic images that result represent the conditions that enabled their creation: her trained physical consciousness and the "secret sympathy" she practices in relation to the natural world. Her movements thus represent an ideal of Nature—an ideal of what the natural world can be and mean for humans; and they represent Nature as an ideal—that is, something in whose creation humans participate, bodily. In this way, then, the kinetic images of Duncan's dance *incarnate* values. They express the value of bodily movement as the activity in which humans create and realize a relationship with what appears to them in and through that movement as divine. *Blue Danube* (re)presents "dance" as *theopraxis*. In doing so, it stirs in audiences a sense of their power and responsibility for bringing into being their experience of (their relation to) what is.

Revolutionary

A similar analysis may be carried out in relation to many of the dances Duncan choreographed in her 27-year career. One of Duncan's solos from 1924, *Revolutionary*, is telling in this regard. In this dance, a figure begins at the back of the stage in the center, and with the music, strides forcefully toward the front of the stage. She sweeps her arms and left leg up toward the sky before plunging downward to a kneeling position, fists drawn and pressing into the floor. Still kneeling she sweeps her arms and torso up and again plunges her fists straight down through the center of her body into the floor, and again sweeps up and plunges down. She slowly stands up and steps backward, not losing eye contact with the audience, before striding forward again and repeating the sweeping and plunging movements. After the second forward thrust, the dancer pauses in the kneeling position, leans forward across her left knee, and casts her gaze across the spectators, calling them with her eyes to attend to their own revolutions. The dancer again retreats to the back of the stage and repeats the forward thrust pattern before ending in a posture of defiance, right fist launched across her body and forward into space.

Duncan made this dance after two years of teaching dance to Russian children under the auspices of the communist revolution. En route to Russia, she was filled with the highest hopes for realizing her vision for a school. After two years of working with hundreds of boys and girls, often with insufficient supplies of food and clothing, she lost confidence in the Bolsheviks, but not in her vision for a world in which justice prevails. The title of her dance cannot not allude to the complex tangle of her experiences there.

Yet, Duncan was a revolutionary in more ways than politically, and she was severely chastised, especially by the American public, for her ties to Russia. Thus, it is likely that in this dance she is also revaluing what it means to be a revolutionary and locating that potency in the act and fact of her *dancing* rather than in politics per se. It is dancing that Duncan credits with actualizing a conception of life that affirms bodily becoming. It is dancing and not the Bolsheviks who will realize an alternative to the Christian

morality that has permeated western politics. It is dancing that will create
not only a new art or a new politics, but a new religion. To be a revolutionary is to dance; to dance is to engage in revolutionary action, to resist the
forces of "Inequality, Injustice and the brutality . . . which had made my
School impossible" (ML 358).

All these revolutionary dimensions of Duncan's life and work weave
through the thrusting, wave-form movements of this dance. Here, to be a
revolutionary is to participate in rhythms of approach and retreat, reaching
to the sky for the highest ideals and plunging with the force of those ideals
into the earth, drawing from the earth the strength and will to generate new
ideals, become those ideals, and again plunge into the earth, and make them
real. Duncan's revolutionary expresses faith in the earth; her revolution
involves wresting the creative potency of bodily becoming from the sway of
ascetic ideals and affirming herself as a creator of value. Echoing Nietzsche
Duncan insists: "every part of life should be practiced as an art . . . the
whole expression of our life must be created through culture and the transformation of intuition and instinct into art" (AD 127).

Revolutionary is generally perceived as characteristic of Duncan's late
dances. The movements are earthy and directed downward. The tone is
heavier, the mood darker, the sentiments more serious. At the same time, this
dance shares a trajectory evident in earlier works: Duncan recreates in herself
the movement impulses she experiences in response to the issues and
ideas that move her, and in doing so, enacts her ideal of a relationship to
those themes that discerns, honors, and realizes a divine continuity coursing
through all of life. Thus Duncan revalues the body-denying values that she
associates (along with Nietzsche) with Christian morality. She enacts the
power and creativity of bodily movement as a medium in and through which
we create and become our highest ideals. In sum, Duncan made dances to
communicate to and effect in others the transformed experience of bodily
becoming that her dancing effected in her. As such, her dances participate in
the revaluation of Christian values regardless of whether the subject matter is
explicitly "religious" or "Christian." She made dances that present "dance" as
an experience open to all in which everyone may learn to love *life*. Duncan's
dance presents "dance" as an aesthetic activity through which life is justified.

Performance

A final dimension of Duncan's dance process (alongside her vision, school,
practice, and dance-making) further illuminates how Duncan intends her
dancing to catalyze a rebirth of religious values. Dance is a performance art; it

lives in the relationship between audience and dancer that the time and space of the dance enables. In this sense, a dance is never present, nor is it ever representable. A dance exists only in and as the movement through which bodies become what they are; it is gone when the curtain falls. What a dance communicates, then, is a transformation of experience occasioned in the space/time of its performance. However, in order for that communication to occur several conditions must be met. A dancer must not only envision, practice, and make dances, he must also establish a relationship with the audience that enables them to receive what he has to offer.

In this regard, Duncan develops a performance aesthetics that reflects the influence on her philosophy of *Birth of Tragedy* as elaborated in chapter 4. As she describes, the primary means by which dancing communicates is *contagion*. Observing a figure from ancient Greece she writes:

> This figure is the best example I could give of an emotion taking entire possession of the body. The head is turned backward—but the movement of the head is not calculated; it is the result of the overwhelming feeling of Dionysiac ecstasy which is portrayed in the entire body. The chord from the lyre is still reverberating through all the members. If you had before you a dancer inspired with this feeling, it would be contagious. You would forget the dancer himself. You would only feel, as he feels, the chord of Dionysiac ecstasy. (AD 131)

Several implications follow from this passage. The first is that the medium through which dance communicates is the body, not the costuming nor the music, not the narrative nor the title. The dancing communicates by virtue of the fact that spectators see human bodies moving.

A second implication follows from the first. The dancing communicates not only because the spectators see bodies moving, but because they themselves are also bodies moving, if less vigorously, during the moment of performance. Seeing the movement catalyzes the kind of visual and visceral identification Nietzsche describes in Attic tragedy. That double identification draws on our instinct for imitating gestures; we can sense another's body as like our own and not in what it looks like but in what it feels. What the spectator "catches" then is not the emotion itself, nor an exact copy of the dancer's experience. What a spectator catches is her own recreation of the emotion that the dancer's kinetic images (recreated impulse–responses) appear to express to her. She must feel what it feels like to make the movement and from that point of connection generate ideas about what that feeling might mean based on her own recreated responses. As Duncan affirms, all gestures—whether seen or made—give rise to an "inner response": "the attitude we assume affects our soul" (AD 103). In short, a spectator can understand the dance not because she is a body by nature but in so far she has learned to sense through her physical consciousness and exercise her kinetic imagination.

A third implication follows from these two. In order for a dance to communicate participation in the experience of bodily becoming, a dancer himself must be moving from an awakened soul. He must have undergone the awakening of his own physical consciousness in order for his movements to be able to communicate participation in such an awakening. As Duncan hymns: "Yes, let us admire the natural dance of the young women and the young men, their transformation, under their pulsating rhythm—in their dance that is perfect and complete, containing all life and bringing the dancers close to the gods. The crowd will follow their steps, will watch their attitudes, becoming one with them in a perfect harmony, combining, on their side, the most noble expression of human life and the clear call of divinity" (AD 98). In the dance, the dancers exercise their trained physical consciousness. They allow themselves to be transformed "under their pulsating rhythm." As the crowd follows, watches, and identifies viscerally with the dancers, they too feel their capacity to make movements that are "godlike"; they feel themselves to be gods. They know themselves, in that moment, as creators and created. The "call of divinity" spectators hear is thus a call to enter into this god-making dynamic—to affirm it and participate in it as integral to the work of becoming who they are: *human*.

These points of Duncan's performance aesthetic hark back to Nietzsche's characterization of art as the only alternative to the ascetic ideal. For him, art realizes an alternative mode of valuation, not by representing it in the form of a competing idea, but by revealing the physiological basis of all of our thoughts, actions, and creations. Duncan follows him here. As she insists, "My Art is just an effort to express the truth of my Being in gesture and movement" (ML 3). This "truth" cannot be confused with a notion of "self" that appears in theories of art as self-expression, or with an "essence" or "I." Art, for Duncan, *expresses* the truth of her "Being" by being the activity in and through which that "Being" takes shape as a physical consciousness of her participation in a universal Dionysian movement. There is no self outside of the dance itself. The self expressed in dance is a process of its own creation through the making and becoming of kinetic images. When Duncan writes that a spectator would "forget the dancer" in seeing the dance, her point is exactly this: what a spectator sees in a dance is the spectator's own recreated image of the dynamic process of self-overcoming the dance represents a human body to be. A spectator sees *dance* being valued as an *expression of life*.

In sum, for Duncan, the transformation a dancer experiences and communicates in the moment of performing is his participation in his own bodily becoming, in his work of awakening and engaging physical consciousness. It is these "feelings" of participating in the constitutive rhythms of the universe that Duncan intends her dances to convey. To stress the importance of a dance's transformative effects, however, is not to deny that the content or theme of a dance is relevant to what it communicates. The point is that the relevance of that content cannot be interpreted literally. The content is relevant because it represents the material that the experience of transformation allows a spectator to perceive anew in the light of his revitalized physical consciousness.

When these conditions hold—when the dancer moves with awakened soul and the spectator actively receives the kinetic images into his sensory field—then what Duncan describes as a "miracle" can happen. A spectator sees *soul*. In generating and incarnating kinetic images of a universal Dionysiac movement, the dance appears as a revelation and a prayer. Of such a dance Duncan writes:

> So pure, so strong, that people would say: it is a soul we see moving . . . Through its human medium we have a satisfying sense of movement, of light and glad things. Through this human medium, the movement of all nature runs also through us, is transmitted to us from the dancer . . . It is a prayer, this dance; each movement reaches in long undulations to the heavens and becomes a part of the eternal rhythm of the spheres. (AD 56–7)

Again, Duncan's religion language in this instance may seem frustratingly opaque and vague until illuminated by her connection to Nietzsche. To see *soul* in a dance is to see a body as the movement of its own creation. It is to see that creative movement as bringing into being the world that is desired. It is to see an ideal of *life* itself as a process of creating—in and through bodily movement—our ideals of beauty and holiness. To those who watch with such expectation, a dancer can appear divine[4]; and the dance can "implant in our lives a harmony that is glowing and pulsing" (AD 103). A dancer dancing can stir hope in our capacity to bring into being a way of life lived accountable to the practice, choreography, and performance of dance as religion.

Overcoming Physiological Contradiction

[D]ancing is the ritual of the religion of physical beauty. The dancer of the future will have to suit the dance to the symmetry of the body. She must have a perfect body, which will again be recognized as beautiful, pure and holy. And in this

body, a free, great spirit must find harmonious utterance in the excitement of the dance. Only in this way can dancing be raised to its place among the fine arts.

—IS 33

Reading Duncan through the lens of her relationship to Nietzsche sheds light on nearly every contentious issue in Duncan scholarship. It does so by revealing apparent contradictions in her claims as traces of her commitment to a project of revaluing Christian values wherever those values occur—politics, fashion, physical education, art, or of course, religion. In each case, moreover, the analysis of Duncan that Nietzsche enables casts new light on his work as well. To conclude this chapter and Part 2, I offer one example—the ideal of woman—and interpret how Duncan's apparently inconsistent statements concerning "woman" represent how she engages and critically advances Nietzsche's project along the trajectory of her engagement with him.

In assessing the stakes of her vision for dance, Duncan is clear: although the liberating effects of dancing pertain to both men and women, those who identify themselves within the Christian West as women have more to gain. In so far as "woman" has been identified with "the body" in opposition to the man with the mind, then the faith in the body that dancing expresses promises to act directly on the ways in which women learn to conceive of themselves, conduct themselves, and relate to their bodily selves. Duncan seems unambiguous.

> It is not only a question of true art, it is a question of ease, of the development of the female sex to beauty and health, of the return to the original strength and to natural movements of woman's body. It is a question of the development of perfect mothers and the birth of healthy and beautiful children. The dancing school of the future is to develop and show the ideal form of woman. (AD 61)

Yet the mention in this passage of an "ideal form" and the "original" and "natural" movement of a woman's "body," leads scholars to conclude that Duncan was, despite her best intentions, mired in essentialism: she recuperated a notion of the body as a material thing whose biological and physiological character determine the fate of a woman's life.[5] Her "woman" is not free.

However, within the context of Duncan's relationship to Nietzsche (and this quotation comes from an essay she wrote just after beginning to read Nietzsche), her appeals appear as attempts to revalue those very conceptions of body and woman for which she is accused. By claiming that "woman" is an "ideal form" she is stating that what it means to be a woman is a product (at least in part) of human creativity. Not only that, she is also asserting that "woman" is a product of the bodily becoming she intends her dancing to exercise. Recall that for Nietzsche as for Duncan, an ideal is a human creation—an expression of value. It is most basically a metaphor rooted in experience that functions in turn to inform experience. By claiming woman as an ideal, then, Duncan is removing the question of what it means to be a woman from the realm of the biological or the religious to the realm of the aesthetic.

At the same time, Duncan does not identify the process of ideal creation with the forces of social construction popular among followers of Foucault, Bordieu, and Derrida. Ideal creation, she maintains with allegiance to Nietzsche, is itself a *physiological* process. It is a process inevitably informed by considerations as simple as food and climate, and as complex as educational, moral, and religious systems. At the same time, as a physiological process, it always remains to some degree accountable to the life force and form of a person. In other words, in line with Nietzsche, Duncan embraces the act of ideal-making as bodily as a way to deny absolute truth to any particular ideal. Rather than espouse either biological determinism or social construction, she holds to a notion of aesthetic justification: any ideal of "woman" must be held accountable by each and every "woman" to the kinds of health its incarnation produces.

Read from this perspective, Duncan's claims are revolutionary within theoretical circles even today: she is arguing that dancing is a medium in and through which people can learn to participate in creating the meaning that their bodies have for them. In dance, as she envisions it, people sanctify their own bodily selves as the creative, reflexive movement in and through which they generate and become (their) gods and goddesses.

Further, if ideal-creation is and should be physiological, and if the bodies of women provide women with valuable resources for discovering movement and unfolding the creative potency of kinetic imagination, then it follows that those who identify themselves as "women" are primarily responsible for generating ideals about what "woman" is. When Duncan claims that dancing is "a wonderful undiscovered inheritance for coming womanhood" (IS 43), she is affirming that the current construction of woman is obsolete, that women themselves must move in order to discover resources for new ideals, and that they must do so based on what they have suffered at the hands of Christian antipathy to their bodies. When Duncan asserts of this woman, "She shall be a sculptor not in clay or marble but in her body . . . In the movement of the body she shall find the secret of right proportion of line and curve and—the art of the dance she will hold as a great well-spring of new life for sculpture, painting and architecture" (IS 43), she is not claiming that all women will find one "secret." She is affirming that no ideal can be true for someone unless it expresses her disciplined attention to the rhythms of her bodily becoming. While there were numerous calls at the time for a "New Woman," Duncan's remained unique: women must dance in order to create ideals that revalue Christian antipathy to female bodies.

Still, the question arises: does Duncan go too far when she claims that the "beauty of woman" is "the guide of human evolution toward the goal of the human race, toward the ideal of the future which dreams of becoming God" (AD 79)? Perhaps. Yet not when read as an instance in which Duncan is overcoming her own physiological contradiction. As Duncan knew from her own experience, the female body has functioned throughout the history of art and philosophy in western civilization as an ideal for beauty. Duncan affirms this fact and its influence on her. It is not a fact she can simply refute by mentally dismissing it. It lives in her, in her body, and in her love for dance. Instead, she works to dismantle its oppressive manifestations, not by giving the ideal of woman's beauty a new content, but by shifting the locus of its authority. By insisting that *women* define for themselves *in dance* what this ideal means, she implies that women must engage in the hard physical work required to develop the physical consciousness from whose perspective they can embrace the idealizing of woman as an occasion to participate in its meaning.

Finally, even if one does embrace the idea that Duncan is claiming for women the right to participate in the making of the ideals that express and encourage love for their own bodily becoming as a source of revelation, it seems that Duncan overdetermines the outcome of the process. As noted earlier, the "perfect body" of a woman, it seems, is a mother (AD 61). Or, as she writes: "And when they become embodiments of the modern vestals, they will be transformed: women in love with love and with the joy of motherhood. At this moment their dance, completed and distinctive, will be the most beautiful of all" (AD 98).

Three points are worth noting here. The first is that Nietzsche himself uses maternal imagery to describe how the creative process is a process of self-overcoming—a process of giving birth to oneself as a mother as one gives birth to the life of another (LaMothe 2003). Duncan may have been doing what revaluing values requires: reclaiming maternal imagery for herself and taking responsibility for assessing its philosophical significance.

The second is that, in so far as maternal imagery has been a primary lever by which Christian authorities have justified the marginalization of women within the power structures of the church, Duncan cannot not engage this imagery and succeed in revaluing Christian attitudes toward women's bodies. She is demanding that our highest and holiest sense of what it means to be a mother be held accountable to our dancing bodies. The implication is radical: to become a "mother" involves becoming a strong, alive, self-creating, free spirit. It means being physically attuned to one's bodily becoming. It requires a discipline of holy selfishness, and a realization of a love for oneself that overcomes the Christian ideal of self-sacrifice.

Third, Duncan was also resisting the call of women in her day, the "feminists," who argued that women should be like men, with the same rights and privileges. Duncan took a position similar to feminists today who assert the difference between "men" and "women." Duncan denied that women should try to conform themselves to the male norms of citizenship or individuality. At the same time, she did not fall off the rope into biological essentialism as she is sometimes accused. The "essence" of "woman" for her *is* to participate in the inherent creativity of a body as it generates the meaning of its sex. For Duncan, her dancing is a medium through which women of all kinds can exercise this creativity. In making this claim she again effectively dislocates any biological basis for discrimination.

This (Nietzsche-enabled) reading of Duncan, in turn, illuminates a trajectory in Nietzsche's work he was unable to complete. It suggests that the revaluation of values as he envisions it cannot happen until and unless those who identify themselves as "women" engage in the process of their own bodily metaphor-making. For women to claim this power of ideal making for themselves is to enact their defiance of Christian morality by expressing faith in (their own) bodily becoming.

That Nietzsche himself predicts this overcoming of his own work is evident in several places, and most especially in the antagonism toward women he demonstrates in his later works. As noted earlier, in Zarathustra's dream of the tombs of his youth, and in Nietzsche's reflections on his mother and sister as the greatest challenges to his self-overcoming, signs of this tension appear. He perceives his mother and sister as challenges to his self-overcoming, not only because they are his relatives, but because they are women, Christian women. Because they incarnate the Christian ideal of woman in relation to man, they constantly evoke in him the Christian man he has been taught to be, the Christian man he is striving to overcome. The implication here is that Nietzsche, in order to overcome the Christian man in himself, needs to be in relationship with a woman seeking to overcome the Christian woman in herself. Nietzsche thought he had found such a person in Lou Salomé. That he mourned the loss of her friendship as a loss for his philosophy attests to his awareness of his own growing edges.

At the end of the Fourth Part of *Zarathustra*, Zarathustra seems poised to learn what Nietzsche's life conditions prevented: how to cultivate solitude as the condition for intimacy with a partner; how to foster a love between equals that finds expression in a desire for a child. Isolated by illness, Nietzsche was unable to learn these lessons. If someone finds in dancing a transforming experience akin to what Nietzsche created through his illness, a rosier future may be possible.

Part 3

Martha Graham

[S]o completely absorbed in this instrument that is vibrant to life.

—BM 15

Sitting on the floor. Soles of the feet pressed together. Hands lightly resting on ankles. Spine stretched, rooted in the ground, reaching through the top of my head towards the sky. Breathing calmly. Waiting.

"And . . . one!" *Contract*! Exhale, deeply, emptying out the inner surfaces of abdomen, curving in and pulling up, expelling all air into infinity. Belly pressing into spine, chin rising. Head and spine hold the arc and curl over to hover above the touching soles.

Pulses—one . . . two . . . three. Sitbones dig into the earth. Pelvis cupping, ribs and neck relaxing, freed by the squeeze of the contraction. With every pulse, the contraction deepens, the breathing pushes out through the skin of the back, opening a suspended space, a space of suspension. A sense of center. And fifteen . . . sixteen . . .

Release! Diving to the bottom of the pulse. Lengthening the spine, tail to top of head. Inhaling, pulling air up the flat front of the body, up through a rising arc. Pulling against the torso from deep in the pelvis. Pausing as the spine peaks, perpendicular to the floor, in a light lifted sitting. Chest expands, dissolving into air. Feet reach forward, out to the wall in front, knees, ankles, and toes along a trajectory of energy and intent. Legs open, sailing across the floor, away from each other, with arms following, trailing overhead.

"And, one!" *Contract* again, deeper into the source. Scooping, emptying, breathing, surging to life. Allowing the form of the breathing to carry me, double me up, head over lower body, an open curved middle, pulsing.

Sinking into self and finding an open middle ground. Remembering it. Knowing it. Knowing myself knowing it. Releasing into deeper and deeper holds. Sixteen more pulses and then . . .

Release! Diving, lengthening, arcing up; stretching, rising, head, neck, and chest floating on breath. Legs move slowly, back together again, pressed straight, toes extending.

And *Contract*! Toes flexing back hard against the belly pressing back so hard as to lift the pelvis and heels off the floor. Shoulders, neck, head, fingers, calm, deny any part in the work. Sixteen pulses again.

Release! As head floats upward, buoyed on cushions of breath, feet curl easily in, around the open lifted middle. Body center is awake.

So begins a class at the Martha Graham School of Contemporary Dance.

Chapter 6

An Affirmation of Life

My dancing is just dancing. It is not an attempt to interpret life in a literary sense. It is the affirmation of life through movement. Its only aim is to impart the sensation of living, to energize the spectator into keener awareness of the vigor, of the mystery, the humor, the variety and the wonder of life; to send the spectator away with a fuller sense of his own potentialities and the power of realizing them, whatever the medium of his activity.

—Graham in Armitage 1937:102–3

In 1924, Duncan choreographed and performed *Mother*.[1] In 1931, Martha Graham choreographed and performed *Lamentation*.[2] Each of these dances features a woman, alone on stage, performing a dance of her own creation for a general audience to compelling solo piano music. Each dance is a portrait of grief, honoring the loss of children or youth, to accident, war, or fate. In each the dancer barely moves. Duncan's figure, standing at first, moves from up stage left to center stage in three sequences of walks, only to kneel and then sit on the floor. Graham's figure spends the entire dance on a bench, center stage. The minimal movements in each piece were so stripped of the pep and shine associated with the popular vaudeville and showgirl entertainers that they barely registered in public perception as "dance."

Against these similarities, the differences stand out. Duncan's "mother" spills from movement to movement with fluid grace. Her arms reach and curl in unending flow; her torso bends and circles, while her lower body remains fiercely rooted in the earth. The soft folds of the dancer's gown move with her cycles, brushing the air. Graham's figure, rooted to her bench, sheathed in a purple tube of fabric, rocks tightly back and forth. Her movements are stark, percussive. Arms reach and punch, stretch and twist against her torso. The body convulses, doubles over, opaque and articulate in its silence.

The interval between these two dances is one scholars often cite as marking a sea change: a new generation of dancers emerged.[3] They were later named "modern" dancers for the ways in which they, like other modern artists of the time, stripped away traditional forms of dance practice and performance, abstracted the "essence" of their art, and created pieces designed to exploit its expressive potential, in this case, bodily movement. Among the modern dancers, Graham in particular has been heralded for distilling Duncan's vision for a dance of the future into a technique and aesthetics of dance training and performance.[4] According to this "great heroine" narrative of dance history, Duncan's religion language, and her claims for dance as religion are considered part of the fluff that Graham pared away to reveal the essence of "dance."[5] In this reading, where Duncan, in *Mother*, subsumes the dancing to the story of a mother's tragic loss, Graham, in *Lamentation*, uses pure movement to express the felt experience of an emotion.

Yet, setting these two dancers in the context of their relationship to Nietzsche's philosophy suggests a contrasting interpretation of these two dances in particular, and of these women's projects in general. Graham, like Duncan, read Nietzsche and was deeply influenced by his vision for dance. In fact, many of the differences between Graham and Duncan may be read in terms of what Graham learned in the wake of Duncan's experience about how to realize a Nietzsche-inspired conviction they shared. Both believed that the task of making good dance (that is, "art") must proceed hand in hand with a revaluation of Christian values. However, where Duncan claimed explicitly that dance is religion and made dances designed to communicate participation in the physical consciousness of our relationship to a universal Dionysian movement, Graham was more circumspect. Graham used religious language in ways that reveal the meaning of that language as dependent on the practice and performance of dance; she made dances that present dance as the activity in and through which the symbols and ideals that comprise our religious lives have meaning for us at all. Thus, while Graham also realizes a vision of dance as what I am calling theopraxis, she does so differently. She enacts in her practice, making, and performance of dance, a contested interdependence of dancing and writing.

Mother and *Lamentation* exemplify the nexus of similarity and difference binding Duncan and Graham in their relation to Nietzsche. In *Mother* the movements of the dancer appear natural, familiar, and graceful. They tell a story. We see a figure awash in currents of grief. In her undulating movements, we see a mother beckoning to her lost children, gathering them in her arms, and letting them go. She appears as embodying a Dionysian paradox: she is lost to the infinite waves of longing that her strong, fluid bodily movements make visible and visceral. She is affirmed as an individual as she recreates the forms of her membership in rhythms of life and death that overwhelm her. Her wave-form movements thus communicate participation in an awakened physical consciousness.

In *Lamentation* there is no story to tell. The figure's movements, sharp and angular, do not appear as natural or familiar. They *appear* to be *stylized*. They call attention to themselves as symbols—as kinetic images—that do not refer to points of a narrative. The figure does not act out grief. Rather, the movements present themselves as representing a process by which people *articulate* their experience, that is, sense, discern, and give meaning to their experience. The movements express a process of creating kinetic images. Said otherwise, what we see in *Lamentation* is not an outward expression of an inner emotion, but, as in the opening quotation, a body's "potentiality and power": its ability to make and become kinetic images of its own experience. We see a body making movement forms that give its inner experience a shape, and thus allow for its release in symbolic form. We see art as Nietzsche's saving sorceress. We see that *bodily becoming* is a person's way of participating in the "sensation of living." Graham's dance reveals a human's sense of her inmost self as a function of her bodily movement, her own kinetic image making.

This interpretation finds further confirmation in the title of this piece. A lament is a song—an elegy or dirge; it is a symbolic expression of grief in and through which people explore and release emotions for the well-being of themselves and their communities. By presenting this dance as a lamentation, Graham implies that *dancing* is the activity that provides the effective meaning of a lament; the act of making kinetic images is the "listening" through which the wailing tones of a song exert their ability to discharge sorrow. Viewed as a lamentation the stylized movements of this dance present a body's ability to create and become (kinetic images of) itself—and not rational justifications, verbal skills, or Dionysian ecstasy—as the potentiality and power that enables humans to *affirm life*.

In short, when read in the context of these women's common commitment to a Nietzsche-inspired vision, the differences between *Mother* and *Lamentation* represent what Graham may have noticed about Duncan's approach to realizing that vision. Specifically, Graham would have witnessed how the religion language Duncan used to describe her vision for dance was misinterpreted time and again, even by those who purported to follow in her footsteps. Duncan's claim that dance is movement flowing from an awakened *soul* was misconstrued to mean that only feeling matters. Her call for *natural* dance revealing the *holiness* and beauty of the body was taken to justify untrained, spontaneous movement. In either case, Duncan's intention to renew the meanings of these words through her practice, teaching, and performance foundered against the stronger forces of cultural habit in thinking about religion and dance (through the lens of the ascetic ideal) as opposing realms of human life, spiritual versus physical.

Evidence in Graham's writing, speaking, and dancing suggests that what Graham learned from Duncan's experience impelled her further along a trajectory they shared. Graham realized that dancing would not be taken seriously as an art or theopraxis in the American context until dancers succeeded in elevating dancing alongside verbal practices of reading and writing as a complementary medium for interpreting and generating the meaning of the symbols, stories, themes, and texts shaping religious experience and expression. Where Duncan danced to communicate participation in a Dionysian experience of an awakened physical consciousness, Graham danced to communicate participation in the process of kinetic image making itself. In this way, Graham sought to accomplish what Duncan (also) intended—redefining religion as a function of bodily becoming—but in her *dances*, rather than in her *words*. In this way Graham's work poses a further challenge to the ascetic ideal: her dances present "dancing" in contested interdependence with textual practices as a complementary medium for interpreting symbols, themes, and texts from western, mostly Christian, religions.

As detailed in this chapter and the following two chapters, the evidence supporting this thesis abounds. It appears most notably in Graham's use of religion language. While Graham refused to discuss religion (claiming that religion is a matter of "personal choice" [Horosko 1991:10]), she nevertheless winds religious vocabulary, symbols, themes, and characters throughout her dance process to an even greater degree than Duncan. Graham uses words and texts of and about religion as stimuli to dance; she treats them as crystallizations of kinetic human experience whose meaning her dancing is uniquely suited to remember. Thus, while Graham does not claim as Duncan does that her dancing *is* religion or religious, she claims recognition for dance given the ways in which dancing revalues the relation of dance to religion determined by the ascetic ideal. In this interpretation of the relationship between Duncan and Graham, religious impulses are integral and even essential to their respective efforts to realize the potency of dance as art.

The rest of chapter 6 explores how Graham's *vision* for dance as an affirmation of life through movement represents her Duncan-informed participation in Nietzsche's revaluation of values. She too seeks to revalue dance as theopraxis. Chapter 7 offers an analysis of the *practice* through which she seeks to realize that vision, and Chapter 8 concludes the discussion of Graham by examining how, in *making* and *performing* her dances, she enacts a conflicted and mutually enabling relationship between the bodily activities of reading and writing, on the one head and dancing, on the other.

Physiological Contradiction, Revisited

To love dance and to want to be a dancer is to enter into a conflicted relationship with Christian values. That this statement was true for Graham as well as for Duncan attests to how little had changed in the generation between them, despite increasing appreciation in Europe and America for modern or contemporary dance as an art. That Graham was aware of this dynamic as driving her work in dance is embedded in the ways that she, like Duncan, acknowledges living a *physiological contradiction*. Graham consciously invokes "religion," Christianity in particular, as both resource and obstacle for her art. Like Duncan, she found in Nietzsche a philosopher who could help her name, understand, and negotiate the conflicted relationship to Christianity that her love of dance expressed. In particular, like Duncan, she found in Nietzsche's work a vision of dance as capable of transforming this conflict into a stimulus for revaluing Christian values toward dance and religion.

That Graham locates this physiological contradiction at the driving heart of her relationship to dance is evident in her descriptions (long after reading Nietzsche) of her "first dance performance" and her "first dancing lesson." At the age of 3, Graham skipped down the aisle of her family's Presbyterian Church in the middle of a service, scandalizing the Pittsburgh congregants, her mother in particular. Whether this first dance performance happened exactly as described is less important than the sense of Graham's recounting: being in sanctuaries where people love, court, and incarnate the divine, she is inspired to dance. Dance is her way of doing what others in that place do: express their highest values. At the same time, as her narration implies, Graham is aware that the dancing she feels impelled to do poses a radical challenge to the expressions and experiences of divinity that others in that sanctuary have learned to value. From this first dance performance Graham learns that her impulse to dance provides her with an empathetic, critical perspective on her Presbyterian heritage as "so without the livingness of the body" (B 37).

Graham's description of her "first dancing lesson" reinforces this sense of engaged critique. The lesson takes the form of a phrase she repeated throughout her career: movement never lies. While many scholars comment on this sentence, few investigate the force of its power for the young Graham. Admittedly, her father said the words to her in an admonishing tone. However, its sting derived from her own faith. As she explains in her *Notebooks*: " 'Movement never lies—' at 4 or 5 was an admonition—'Lie' in a Presbyterian household was and still is a clanging word which sets whispering all the little fluttering guilts which seek to become consumed in the flame of one's conscience" (N 270). The ongoing force of the statement for her hinges on the fact that "lie" "was and still is" a "clanging word." To lie is to sin—it is to commit a terrible act, an act to be avoided at all costs. That "movement" *never* lies immediately elevates it in the mind of one so formed.[6] In each of these "firsts," then, the thrust of Graham's account traces a physiological contradiction familiar to readers of Nietzsche and Duncan. A deep embodied affinity for the Protestant Christianity of her youth stirs and funds a love for dance that pits her against the Christian values, institutions, and authorities who forbid it.

At the age of 14, Graham moved from the then dark and sooty coal town of Pittsburgh to the sun-filled, religiously-diverse climate of Santa Barbara, California, and in the process developed metaphors for her physiological contradiction akin to those used by Duncan and Nietzsche. As she recounts: "No child can develop as a real Puritan in a semi-tropical climate. California swung me in the direction of paganism, though years were to pass before I was fully emancipated" (McDonaugh 1973:13). Puritan elements remained. She vividly remembers the minister of the Presbyterian Church there who responded to her assertion, "I am going to be a dancer," by turning his back and walking away. Yet, the light, the air, the space; the luxuriant vegetation and bright skies; the conglomeration of cultures, traditions, practices, and symbols encouraged Graham to develop the perspective on her Christian inheritance that dancing afforded her. Graham herself became a Puritan pagan, or pagan Puritan—someone whose regard for Christian religion found expression in an antipathy toward its life-denying forms.[7]

The experiences she remembers having in Santa Barbara were ones that provided her with further insights into this contradiction. Graham relates, for example, the experience of attending mass with her Catholic nanny, Lizzie, at the Franciscan mission. As she notes, "Something of the sunlight and the color of the stone permeated the service and gave it a kind of living-ness" (B 49). That she saw *something* of the sunlight, and a *kind* of livingness gave her a clue. As she came to believe, the anti-dance attitudes of Christian authorities were not the fault of "religion" per se, or even of Christianity. The problem, rather, was an ignorance of dance perpetuated by western culture's faith in verbal acts—in the *words* of a *service* rather than in sunlight and movement—as best able to convey the Christian message. In response she came to believe that those who feel an undeniable pull to dance are responsible for demonstrating what they know: that dancing can *enact* (both represent and realize) the highest values of human living. As Graham later confirmed, "Dancing is movement made divinely significant" (G32, 5).

It was the experience of seeing Ruth St. Denis perform in 1911 that confirmed for Graham that she was such a dancer. She recalls, St. Denis "was a goddess figure. I knew at that moment that I was going to be a dancer" (B 56). Five years later, when Graham enrolled at the Denishawn School, founded by St. Denis and her husband Ted Shawn, her attraction to St. Denis blossomed into reverence: "I worshipped everything about Miss Ruth—how she walked, how she danced. Miss Ruth was everything to me" (B 68). In St. Denis, Graham saw a spiritual being—someone who went to church, read to the dancers every afternoon "in a luminous sari . . . from Mary Baker Eddy and the *Christian Science Monitor*" (B 63), exuded wisdom, and who, when she danced, *was* a goddess (B 70).

St. Denis would only have encouraged the young Graham's perception of her and her work. As St. Denis insisted throughout her career, all of her dances were religious, regardless of subject matter. For her, the key was *intention*, the dancer's state of mind (St. Denis 1916). The act of dancing, for St. Denis, provides humans with a medium for experiencing and expressing their relation to divinity. In her dances, St. Denis intended to become the goddesses whose images she recreated, whether Radha, Ishtar, Kwan Yin, Isis, or later in her career, Mary, Mother of God.[8]

In confirming her sense of her own task then, Graham's seven years of studying, teaching, and performing with the Denishawn Company proved invaluable. In a letter to Ted Shawn in 1923, Graham wrote: "I don't believe you know how truly my heart is in Denishawn—because Denishawn—in all—is my religion—What it is built upon is what I believe." She continues: "I was born there you know. The artist in me was born in Denishawn" (G23). As these sentences suggest, Graham's time with Denishawn not only seasoned her into a dance professional; it gave her the inspiration and confidence she needed to reweave the strands of the Denishawn vision into her own philosophy of religion and dance.

Critical to this reweaving was another influence at work upon her during the Denishawn years: the accompanist for the company, Louis Horst. It was Horst who first urged Graham to read Nietzsche as well as other German philosophers and modern artists. In long train rides across the country they read together, in translation or with Horst translating; they discussed their ideas. From early on Horst was convinced: Graham would catalyze the birth of an American, modern dance, a dance capable of doing what Nietzsche proposed—communicating participation in an affirmation of life.

As Graham describes it, she left the Denishawn Company for money. She had not been cast in an upcoming Asian tour, and had been offered a lucrative role with the *Greenwich Village Follies* of 1923. She would perform in an oriental-style theater piece called "The Garden of Kama," and two (or three) other solos. She accepted. With the *Follies* she toured the vaudeville circuit throughout the United States and Europe, performing up to five shows a night. She had billing and a high salary; publicity and a following. Yet after two years, Graham decided to leave. In her autobiography she recalls why: "I was not about my Father's business." She continues, "That, of course, is biblical, but I did believe that if you followed what you believed to be beautiful, you would achieve what you were supposed to achieve . . . I wanted to create my own dances, on my own body" (B 103).

This short description, albeit years after the fact, both suggests the influence Nietzsche had on her decision to leave the Follies and encapsulates the use of religion language that marks her post-Denishawn career. In this passage, Graham does not claim to have heard the voice of God. Nor does she claim to be Christian or even to be religious (though she may or may not have been). She uses religion language self-consciously; she admits that her description of her decision is "biblical." In making this self-reflective move, Graham implies that she is _choosing_ to use religion language as a _metaphor_ for describing her relationship to dance. The implication is that religion language itself is and has value as metaphor—as an aesthetic human creation. Moreover, in this instance Graham _chooses_ religion language to describe an experience where she experiences having _no_ choice. The pull she describes as her "Father's business"—the pull to dance—is a pull that leads her to act against reason, to reject the money and fame she was accruing, and to set herself in a collision course with those who reference the Bible, tradition, or Christian history in denying dance a role in religious life. In short, Graham's use of religion language reveals her strategy for negotiating the physiological contradiction Nietzsche and Duncan also acknowledge. By framing her call to create dances on her own body as her "Father's business," Graham identifies her work as a site where an historical Christian antipathy toward dance is overcoming itself.

Branching off on her own, Graham was soon joined by Horst. For the next ten years, Horst served as her accompanist, composer, lover, friend, and champion. Together, they continued their studies of philosophy and religion, of modern and primitive art, and forged a philosophy of dance— an approach to imagining, practicing, making, and performing dance that came to characterize what Graham called her "way of doing things." During these years, Nietzsche's philosophy loomed as particularly important. While there are obvious indications of this development—as when Graham announces "I owe all that I am to Nietzsche and to Schopenhauer" (Stodelle 1984:38)—there are other more subtle indications in her early dances as well that suggest how deeply Nietzsche's influence was informing her work.[9]

Going Solo: *Heretic and Dance*

Early fruits of Horst and Graham's Nietzsche-infused collaboration appear in a piece often cited as distinguishing Graham as a dance-maker in her own right: *Heretic* (1929).[10] Much is made of the contrast in this dance between the lone individual, dressed in white, arrayed against the members of the anonymous group, clad in black. The group forms and reforms itself in strong, percussive unison as a wall against which the individual hurls herself unsuccessfully. At the end of the dance the individual falls to the ground, depleted by her efforts to resist the mass. Most commentators stress the bleak political implications of the dance, setting it in the context of social and political movements of the time.[11]

Yet, a comparison of this dance with Nietzsche's account of tragedy suggests that Graham may have had other ideas in mind as well—ideas that make sense of why she names the piece *Heretic* and not "rebel." The Heretic in white is an image of innocence. She is a dancer, holy in her dancing. At the same time, it is her dancing, her bodily movements, that provokes the heel-thumping, fist-clenched response of the group. This dynamic documents Graham's experience: a dancer who feels herself called to do her "Father's business" finds herself opposed by Christian congregations, and the values, traditions, and history they represent.

Is she doomed? While a literal interpretation of the dance as a narrative suggests as much, a comparison with tragedy as Nietzsche describes it suggests that she is not. The configuration of the piece into individual and group echoes the distinction between the hero and the chorus; and the dancing draws the audience to identify viscerally and visually *with both*. On the one hand, the spectator sympathizes with the dancer, as the individual, the image of a person. On the other, the elemental rhythms of the group are compelling. We, like them, are watching her. "We" are encouraged to see her as a *heretic*, as a challenge to our highest ideals. The effect of this double identification is that in her fall, we are asked to take responsibility for our role in being the wall against which the dancer in each of us breaks. We are asked to take responsibility for participating in a culture that not only demands conformity to its notion of "individuality" but denies dance a role in the religious landscape. As we identify with the heretic and group, our relationship to dance is transformed: we experience the persecution of dance as a persecution of our own bodily selves, a denial of our own kinesthetic awareness. We see how what we are taught to do in our culture is to march in lock step unison.

Read from this perspective, *Heretic* is about dance: it reveals *dance* as the medium in which religious, historical, and personal resistance to dance may be examined *and* revalued as a stimulus to making dances that catalyze awareness in spectators of the physical contradiction they themselves embody—as both lovers of dance and part and parcel of the resistant wall. The juxtaposition of individual and group in this dance calls audience members to account for their role in perpetuating a worldview that denies dance a role in the experience and expression of religion. It calls them to remember their bodily creativity, the freedom they enjoy in the ability to make kinetic images. What world are you bringing into being through your own moves?

More explicit evidence of Nietzsche's influence on Graham's emerging practice of dance appears in a second dance from 1929 titled simply, "Dance." This piece was accompanied in the program with a line from the _Genealogy of Morals_ discussed in chapter 3: "strong, free, joyous action." Such a quotation primes the audience to see this action.

Yet, the riddle posed by this solo is that Graham barely moves. She remains rooted in one place, sitting, and concentrates all of her action into percussive shifts of her torso. As Margaret Lloyd describes: "Martha's first solos were axial. In her attempts to simplify movement to its core, she scarcely moved at all" (Lloyd 1949:50). As Elizabeth Kendall writes, the movement was "pressed back into the torso of the dancer, instilling her with power and intensity within, further compressed by the illusion of something stopping it from the outside. She was ruthless in her experimental reduction of the movement" (Kendall 1979:206). Kendall notes the irony of pairing such compressed movement with the Nietzsche quotation.

Alternately, it is indeed the link with Nietzsche represented by Graham's choice of quotations that illuminates the significance of her compressed movement. In the _Genealogy_ (as described in chapter 3) Nietzsche distinguishes between different modes of valuation so as to call attention to the dynamics of revaluation at work in the Jewish and Christian moralities of western societies. In his account, "dance" is what the nobles do; it is a means by which they preserve their freedom: that is, the physical vitality that enables them to digest their experiences, create their own values, love themselves and their enemies, and in general resist the seduction of ascetic ideals. Dance is also what impotent priests must forbid their flocks as beneath them. In choosing this quotation for her dance, then, Graham situates her art at the heart of this war between competing value systems that, as Nietzsche admits in the Third Essay, "we" moderns suffer as a physiological contradiction. Yet Nietzsche is vague about what kind of dancing the nobles practice. Graham seems to be offering an answer. Her "Dance" _is_ strong, free, and joyous action. But how?

An answer emerges when viewed in the context of Nietzsche's critique of Christian values. Graham's _Dance_ represents what it is about _dance_ that enables a dancer to embody an alternative mode of valuation than that offered by Christian idealism and its antipathy to the body. It is not the steps. Nor is it the intention of the dancer, as it was for St. Denis. Nor is it the harmony of the movement with bodily form, as it was for Duncan. Dancing embodies an alternate mode of valuation by being the medium in which we resist and recreate our relationship to our bodily being.

Dance appears as the activity in and through which we can transform our experience of resisting dance into a source for dance movements that preserve the conditions of our freedom. The very internal and external resistance against which the dancer must move is what makes her strong and free; it is what enables her, forces her, to go deep within her bodily self and find a source of movement that does not sustain the illusions perpetuated by the ascetic ideal. A source of movement that is not mental or "spiritual" but bodily. Finally (as discussed in chapter 7) this dance suggests that Graham finds the source of such revaluing movement not in a physical location (as Duncan did with respect to the solar plexus) but rather in a rhythmic, life-sustaining movement: the pulse of breathing. In short, what we see in "Dance" is how the act of dancing is capable of transforming an experience of physiological contradiction into a reason to affirm life.

A Vision of Affirmation

Perhaps the most significant trace of Nietzsche's influence on Graham appears in a phrase she repeated from the mid-1930s until her death: dance is *an affirmation of life through movement*.[12] In embracing *affirmation* as the aim of her dancing, Graham not only enlists Nietzsche's help in resisting the hegemony of the literary over the sensory in our experience of art, she frames her own work as participating in his project of revaluing Christian values.

As the quotation heading this chapter suggests, "affirmation" for Graham represents the difference between dancing and an interpretation of life "in a literary sense." A dance that communicates affirmation does not deliver an idea about life to our understanding, nor a moral to our intellect. In line with Nietzsche's vision, such a dance imparts *sensation*: it *energizes* a spectator into awareness, opens a *fuller sense of potentiality and power*, and thus reeducates a person to the process of bodily becoming—the process in and through which he or she generates and realizes his or her highest ideals. Such dance, for Graham, is "significant movement" (Armitage 1937:98); it is "movement made divinely significant" (G32, 5); it is "good dance" (G53, 6).

In fleshing out what this idea of affirming life means for her dance process, Graham represents what she has learned from Duncan regarding the challenges of advancing Nietzsche's vision. Graham does not aim to "be the chorus" in relation to other arts as Duncan did. She does not leave the verbal arts in particular so much room to maneuver. Rather, Graham engages texts of and about religion to play an active role in each stage of her dance process; and she does so in order to disclose how the verbal arts depend on the primary level of metaphor-making that dance enacts for their ability to conduct meaning. She does so to revalue dance as *theopraxis*—a bodily activity in which we create and become our highest ideals. In what follows, I trace this dynamic through three aspects of Graham's vision for dance: "good" dance reveals a human being, is of America, and expresses an inner responsiveness of mind and body.

Revealing a Human Being

For Graham, a dance that can catalyze an affirmation of life is one that, first of all, reveals a human being. As she writes: "It has not been my aim to evolve or discover a new method of dance training, but rather to dance significantly. To dance significantly means 'through the medium of discipline and by means of a sensitive, strong instrument, to bring into focus unhackneyed movement: a human being' " (G41, 178). On the one hand, this idea of bringing a human being "into focus" suggests Graham's kinship with other modern artists. She was isolating the constitutive elements of her art: "Each art has an instrument and a medium. The instrument of the dance is the human body; the medium is movement" (G41, 178).[13] Like modern artists as well, she was using this focus on the constitutive elements of her art as a basis for rejecting forms of dance that do not. So Graham distanced her work from what Horst called the "overblown superficiality" of nineteenth century romantic art in which "Dancers aspired to skim and spin like heavenly puff-balls" (Horst 1961:52–3); from the "weakling exoticism of a transplanted orientalism" evident in Denishawn, and from the recreations of wind, water, and trees common among imitators of Duncan (Jowitt 1988:161). By attending to the body *as body*, Graham sought to discover kinetic images that were more concentrated in their execution, more potent in their visual, visceral impact, and more relevant to the human experience of living than those she saw elsewhere at the time. She sought to discover what a human body can do. For Graham, dance is *significant* when it concerns the humans *we are* (G35, 56–7).

At the same time, the image of "human being" Graham sought to bring into focus bears striking resemblance to the image of bodily becoming that Nietzsche promotes. A constant theme in Graham's speaking and writing is one that Zarathustra taught: "One has to become what one is" (Armitage 1937:101). It is the "essence" of human being to change, to die "daily small deaths," and again be "born to the instant" (B 4, 17). A "human" *is* this participation in the rhythms of her own bodily becoming. Moreover, for Graham as for Nietzsche, this process of ongoing recreation is physical though never merely so. A body is "the instrument by which all the primaries of life are made manifest. It holds in its memory all matters of life and death and love" (B 4); and in the act of manifesting each of these primaries, a body creates beyond itself. A body *is* a constant process of generating images thoughts, feelings, and experiences, spiritual, emotional, intellectual or otherwise—of its sensory immersion in the moment. For Graham, then, a dance that reveals a human being is one that reveals a human *as* this bodily image-making process. In so far as a human being is *a movement of bodily becoming*, then a dance fails to reveal "a human" if it only offers an interpretation of life in a "literary sense." Such a dance necessarily fixes what a body is in words, and, as Nietzsche predicts, fails to play its role as the healing sorceress seducing us to (participation is bodily) *life*.

Thus when Graham envisions dance as *significant* movement—movement *made divinely* significant—she is referring to movement that enacts this process of bodily becoming that reveals a body *as* this process of becoming and inspires her to *become* what she is though the ongoing process of her own bodily image making. Such reflexive movement is *divinely* significant in the sense that a person who invests her attention in her bodily becoming, according to Graham, will realize her potential and power as a god-former, a maker of ideals, and come to view herself as godlike. As Graham writes: "I believe so completely in the life of a human being and the sanctity of a human being . . . You stand and you are an individual and you are beautiful. That, to me, is very sacred" (B 72–3). For Graham it is participation in the process of bodily becoming and not a particular experience or belief that realizes the sacred character of a human being. Religious experience or doctrine may represent one possible manifestation of bodily becoming, but not its generative source. In so far as we believe in one such phenomenon as the locus of our divinity, we perpetuate the ascetic ideal by denying our sensory experience and immuring ourselves in idea-edifices of our own creation.

Finally, Graham also shares with Nietzsche a sense of how art, by revealing a human as bodily becoming, works its healing magic: art educates us to our senses, exercises our instincts, and stirs our energy for life. As in the opening quotation, dance that is an "affirmation of life" is dance that imparts the "sensation" of living. Elsewhere Graham elaborates: "In life, heightened nerve sensitivity produces that concentration on the instant which is true living. In dance, this sensitivity produces action timed to the present moment. It is the result of a technique for revelation of experience" (G41, 180). In this account, dance is effective in revealing a human as a bodily becoming because the act of dancing itself exercises the capacity that enables "true living." To learn to dance is to learn to inhabit more fully the scope of sensations present to us in any moment; it is to learn to live through this heightened nerve sensitivity in any moment and learn through those sensations who we are, that is, the movement of our own bodily becoming. Dance, in this regard, exercises and strengthens both an awareness of this dynamic and a capacity to participate consciously in it. When conceived and practiced in this way, Graham contends, dancing is a moment of "passionate, completely disciplined action" that can "communicate participation to the nerves, the skin, the structure of the spectator" (Armitage 1937:86). What is important to Graham about a dance is that people *feel it*, for feeling quickens sensitivity and opens "doors that have not been opened before" (B 202).[14]

In sum, dance as Graham envisions it can reveal a human being by communicating participation in an experience of bodily becoming. By educating us to our senses, dancing resists the cultural forces Nietzsche describes in terms of desensualization and oversensitization; it provides us with a knowledge of ourselves as kinetic-image makers that our systems of education, religious and otherwise, largely deny. Through such education to our senses dance catalyzes a magic transformation, providing us with a physiological perspective from which to affirm, resist, and overcome the workings of the ascetic ideal in us. As Graham confirms, by communicating "experience by means of action and perceived by action" a dance speaks "to that insight in man which would elevate him to a new strength through an heightened sense of awareness" (Armitage 1937:84–5). Again, that insight is a physical consciousness of ourselves as kinetic-image makers; an insight that dance is theopraxis; an insight though which we define what it means to become *human* and thus realize our divinity, our creativity. In sum, dancing as Graham envisions it can grant us a "new strength" to resist the hegemony of interpretations that render life "in a literary sense." It can serve to free us from our idea edifices, and reawaken us to the primary levels of metaphor- making and the inherent creativity of our bodily becoming.

Of America

It is because Graham believes that significant dance educates us to our senses by communicating participation in the experience of living that Graham was so adamant about making dances that are "of America." A dance must not only educate people to the senses in general but educate them to the rhythms that comprise their own time and place and inform their own bodily becoming. Differences in rhythm distinguish one time and place from another, one person from another. In so far as these rhythms comprise the material of the bodies with which dancers work, then a dance that seeks to reveal the human and affirm life must respect and recreate those rhythms. While Duncan too was interested in educating bodily sensation to the rhythms that constitute human life, her attention gravitated to the rhythms of a universal Dionysian movement, or divine continuity.

Graham, by contrast, directs her focus on rhythms to the social, cultural, and political life, the landscape and habitats characteristic of America.[15] As Graham writes: "Granted that rhythm be the sum total of one's experience, then the dance form of America will of necessity differ greatly from that of any other country. So far the dance derived or transplanted has retarded our creative growth" (Armitage 1937:98). Even so, a dance that "differs greatly" in this way, Graham maintains, will be "of America" but not "American" per se. It will express an awareness of its time and place, and in so doing express something that holds for all human bodies regardless of time and place: creative participation in the rhythms that constitute the "sum total" of their experience. As Graham confirms: "It is not to establish *something American* that we are striving, but to create a form and expression that will have for us integrity and creative force" (Armitage 1937:98). Again, rather than present her dance as communicating a particular (for example, Dionysian) rhythm, Graham presented dances that enact a process of kinetic image making—of sensing and recreating rhythmic forms. In so doing, as argued in chapter 8, Graham was making the case for dance as an "American Document," capable of serving a function we generally attribute to verbal practices in representing who we are.[16]

Expressing an Inner Responsiveness of Mind and Body

In an effort to attune to the rhythms that constitute the time and space of America, and create dances that reveal a human being, Graham studied the dance and religion of native people who had lived on American soil before the colonists settled.[17] While Duncan turned to nature and Greek art for resources in developing her wave-form aesthetic, Graham studied tribes in the American Southwest. In doing so, she, like Duncan, perceived dances that at least to her manifest the kind of sensory vitality she intended her dancing to communicate.

That Graham warmed to a fascination with so-called primitive cultures common among other modern artists is not surprising. Duncan and St. Denis themselves provided her with examples of learning from other cultures about how to make dances that would overcome western Christian hostility toward dance as religion. Like them as well, Graham did not approach other cultures with the intent of recreating those cultures' dances; nor did she advocate a return to primitive social forms. Horst and Graham were aware that any attempt to copy native dancing would produce "affectation" (Horst 1961:57). For them, what Horst called their "back to the primitive urge," represented a "going back to . . . go forward" (Horst 1961:53). It represented a strategy for cutting through the sterility and excesses they perceived in late romanticism, and discovering what a body can do. They were looking for models of dancing where the dancing was, "as strong as life itself" (G35, 58), that is, rooted in its time and space in such as way as to reveal a human as the process of its own bodily becoming.

What Graham and Horst found in their studies of native cultures was akin to what Duncan and Nietzsche discerned in Greek art: an example of a culture that embraced bodily movement as an effective symbol of its highest values. Graham describes her responses to a Pine Dance she observed at the Santo Domingo Pueblo thus:

> It was difficult to describe—but nothing I have ever seen or dreamed of equaled that great communal dance in earnestness, intensity, faith in the eternal recurrence of natural phenomena—such savage ruthless awareness of life. It was the most pure holy ceremony. There was not only great soul—but the sense of form was also moving. They are so wise and such great artists.
> (Unpublished letter, September 7, 1931, Helpern 1994:6)

In this account, what impresses Graham about the ceremony is a "ruthless awareness of" life paired with a courageous willingness to affirm it—to have in a Nietzschean phrase, "faith in the eternal recurrence of natural phenomena." As an expression of such faith, Graham concludes, the dance was "the most pure holy ceremony," and "holy" for reasons her work with Nietzsche would have inclined her to recognize: it facilitated an enabling, encouraging encounter with (the abysmal thought of) eternal recurrence. The dance seemed to Graham to accomplish this task, moreover, by enacting an awareness of the body, of bodily becoming, as the creative medium in and through which humans know (ourselves as) elemental rhythms. "The sense of form was also moving." For their ability to move their bodies in ways that communicate and effect this affirmation of life, Graham concludes, "They are so wise and such great artists." The following summer, Graham spent her Guggenheim Fellowship—the first for a dancer—traveling in Mexico, studying the art and architecture of Mayan and Aztec civilizations.

In her studies of these cultures, Graham distilled lessons for her own dance process that her experience of Duncan and Nietzsche's work helped her perceive. Keen to discern the "secret" of these dances, Graham zeroed in on what Horst called the "inner physical, muscular consciousness which colors movement and gives it its particular quality" (Horst 1961:48). As he confirms, the Native Americans' "ability to derive dance movement from life experience rather than from a sort of decorative habit" seemed to indicate a "deep responsiveness between body and mind" that formed the self-creating substance of their art (Horst 1961:59). It was a secret, Horst admits, that he and Graham credited Duncan for discovering as well in the name of making

"religious" dances whose movements flowed from an awakened "soul."
As Horst writes:

> The pioneers in modern dance and their successors recaptured . . . an inti-
> macy with the muscle tensions of daily movements which had been
> lost . . . This is . . . an inner sensitivity to every one of the body's parts, to the
> power of its whole, and to the space in which it carves designs. The great
> quest was to find ways to attain this sensitivity, and manners in which to
> discipline it for communication. (Horst 1961:17)

For Graham and Horst, the significance of this discovery loomed large:
it provided the key for generating an approach to dance practice, making, and
performance that would participate in the revaluation of values Nietzsche
describes along the trajectory opened by Duncan by setting dance on its
own feet as a medium for generating ideas of and about religion. In so far as
the absence of an inner, physical consciousness testifies to the power of ascetic ideals at work, then any attempt to cultivate such physical consciousness
and exercise it in creating kinetic images of human being is necessarily
engaged in a project of revaluation. As Graham phrases it, she intends her
dancing to revalue the "puritanical concept of life," underlying the core
values on which western societies have been built, including those values
that sustain a perception of religion and dance as mutually exclusive and
often hostile activities. In so far as dancing educates us to our senses, to our
bodily becoming, it provides us with a critical perspective from which to
assess the value of our values, and in particular, to revalue conceptions of
art and religion that deny the body a contributing role. As Graham writes:
"living is an adventure, a form of evolvement which demands the greatest
sensitivity to accomplish it with grace, dignity, and efficiency. The puritanical
concept of life has always ignored the fact that the nervous system and
the body as well as the mind are involved in experience, and art cannot be
experienced except by one's entire being" (G41, 180). Dance that cultivates
a deep responsiveness between mind and body itself incarnates new
values—including a value accorded to dance as theopraxis.

At the same time, Graham moves farther along the trajectory of revaluing
religious values than the "pioneers." Where Duncan relied heavily on her
studies of nature and Greek traditions in discovering what a body can do,
Graham focuses her studies on the movements of her body and those of her
dancers. Where Duncan cultivated a bodily awareness of movement
impulses arising in the solar plexus, Graham isolates and stylizes the rhythmic
movements that comprise bodily becoming itself—the rhythms of breathing. As such the sensation of living Graham seeks to impart is the experience
of being the self-creating kinetic-image maker who can awaken within herself,

as one possibility among others, a physical consciousness of herself as a soul in relation to a universal Dionysian movement, and can recreate the forms of that consciousness in kinetic images that communicate participation in the experience of it. As Horst confirmed: "Of this deep responsiveness between body and mind the art of the dance is formed" (Horst 1961:14). In this way, Graham critically advances Duncan's project by representing Duncan's accomplishment as what dance is without binding herself to the particular form of its expression that Duncan espouses.

This difference, while it may seem minimal, not only carries substantial implications for practicing, making, and performing "dance," it also provides Graham with a ground for establishing dance alongside verbal practices as a realm of symbolic expression in its own right—and one that allows us to honor in a way that Christian words have not, the bodily source of life. Articulating this revaluation Graham confirms: "A body is sacred, and should be treated with honor, and with joy and with fear as well. But always, though, with blessing" (B 7).

Primitive Mysteries

Primitive Mysteries is a dance in whose movements these themes of affirmation and revaluation, the primitive and the American, dance and religion entwine.[18] We often hear about the inspiration for this dance (the Native American and Christian ceremonies Graham observed); the process of its creation (taking over a year to perfect the walk used by dancers to enter and exit the stage); and the ecstatic reception of the piece (over 22 curtain calls at the first performance).[19] In the context of this analysis, another interpretation appears: this piece positions dance as an active medium for revaluing Christian values and catalyzing affirmation of life.

That this piece engages Christian symbolism is evident in the three-part structure of the dance; in the names of those parts ("Hymn to the Virgin," "Crucifixus," "Hosanna"), and in movements that allude to the life of Jesus. At the same time, the dance offers a twist: its central figure is a woman in white, Mary, mother of Jesus. The group encircles her, worships her, and blesses her. The dance retells the biblical story, in other words, as the story of Mary—as a story of her conception, loss, and journey toward affirmation. Mary's dancing appears as the locus of what Christians know as the doctrines of incarnation, crucifixion, and resurrection. Mary's body, in its self-creating capacity, appears as that which enables the Christian Word to have meaning.[20]

Nevertheless, Graham does not claim that this dance is or intends to be religious. Her dancers describe their experience of performing it as religious; audience members describe their experience of watching it as religious; but Graham never does.[21] Nor does she describe or justify the dance as an artistic recreation of the forms of a Christian ceremony. Rather, Graham leaves it to the dancing itself to communicate participation in the lives of some of the primary symbols through which many of those raised in Christian culture experience life, death, and rebirth. The significance of this move is all the greater for Graham's linguistic restraint. What "speaks" is the dance, and what it communicates is how powerful dance can be for generating the meaning of "religion." The dance is so, moreover, by virtue of what we, in the "modern" West share with those we deem "primitive": our participation in a *mystery*, in the dawning of a new culture. "For we, as a nation, are primitive also—primitive in the sense that we are forming a new culture" (G32, 6). The effect of this dance, then, is to awaken people to their own "primitive" state vis-à-vis religion; to sensitize them to their bodily participation in the meaning of incarnation, crucifixion, and resurrection, and to energize them to take responsibility for the ways in which they assume the truth of cultural forms which are "of America." This dance implies that the process of generating a new culture and its religion requires that we learn to dance, and that we learn from dance how to embrace our own bodily creativity as integral to that process. It is in and through dance (as Graham envisions it) that we will exercise our god-forming instinct in ways that affirm through movement the primitive mysteries of human living.

Good Dance, Significant Movement

In tracing the various threads of Graham's vision with the help of Nietzsche and Duncan, what appears is a complex, coherent view of what makes dances "good" or "significant." "Good" dances overcome the physiological contradiction that we as a modern culture are living as people who both embody and resist ascetic ideals, and good dances do so by revaluing values. As Graham confirms: "Great art never ignores human values. Therein lie its roots. This is why the forms change" (Armitage 1937:84). The forms of art change to express new values, and Graham's art incarnates the trajectory Duncan and Nietzsche exposed in which Christian values are overcoming themselves.

Specifically, Graham envisions a dance that realizes the overflowing love that Zarathustra teaches. She intends her dance to express a "love of the body" (B 240–1) by enacting a way of cultivating bodily sense as the medium through which one lives, creates, and justifies existence. Her dance represents a use of the body that refuses to cultivate a sense of a dualism between soul and body as a protection against life's most harrowing sorrows. It is a use of the body that honors the body as granting a unique knowledge of its time. It is a use of the body that incarnates health-enabling, life-affirming values. It is a use of the body that acknowledges dance as theopraxis. As Graham confirms: "Dance is an absolute. It is not knowledge about something, but is knowledge itself . . . It is independent of service to an idea, but is of such highly organized activity that it can produce idea" (Morgan 1941:Preface). This idea-producing knowledge that Graham describes is knowledge of how to participate bodily in the process of discovering and becoming who one is. Given this vision, what a dancer does in his practice, as chapter 7 explores, is crucial—not as a preparation for making perfect steps, but for generating the content of whatever his movement will communicate.

Chapter 7

Athletes of God

I am a dancer. I believe that we learn by practice. Whether it means to learn to dance by practicing dancing or to learn to live by practicing living, the principles are the same. In each it is the performance of a dedicated precise set of acts, physical or intellectual, from which comes shape of achievement, a sense of one's being, a satisfaction of spirit. One becomes in some area an athlete of God.

—B 3

As Graham began to make the kinds of dances she wanted to dance, she was, at least at first, not interested in developing a "technique." Like Duncan, Graham understood "technique" as referring to a codified set of formal shapes and exercises imposed upon the body. Like Duncan, she rejected all such techniques for suffocating the subtle sense of inner responsiveness and bodily creativity that she sought to strengthen as the source of dance. Like Duncan as well, Graham's guide word was *discovery*: "No one invents movement; movement is discovered" (Rogers 1941:182). Graham wanted to discover what a body can do and then practice what she had discovered in order to generate new movement material. Graham's early rejection of technique, in other words, like Duncan's, expressed a faith in the body. It expressed her desire and willingness to listen to her body and trust it to guide her in finding movement capable of revealing a human and communicating participation in an affirmation of life—movement that would be *divinely significant*. In this project, the classes Graham taught were her laboratory; her own body and those of her students were her experimental media.

The classes Graham taught in the late 20s and early 30s featured exercises familiar to Duncan as well—the activities of human living which every healthy body does, however idiosyncratically. In addition to walking, skipping, and running, Graham explored the movements of coughing, laughing, and sobbing. Also concerned with finding the impulse of significant movement—the crater of motor power—Graham honed in on the muscular forms of these ordinary movements, and studied what a body does to make them. Over a period of time she focused her investigations on a source different from Duncan. Rather than isolate an anatomical location—namely, the solar plexus—Graham isolated a physiological movement that enables and connects all moments of human living: the exhalation and inhalation of the breath. Graham stylized the movements of breathing into kinetic images she called "contraction" and "release."

While there are several hypotheses (discussed further) about why Graham chose the movements of breathing as the generative kernel for what eventually became the "Graham technique," the one offered in this chapter draws upon Nietzsche's work and Duncan's use of him in interpreting the significance of this choice. In a nutshell, every time a dancer performs a contraction and release, regardless of the narrative context or thematic material of the dance, he enacts the capacity of his body to *make images of his self*—kinetic images. As such, practicing contraction and release expresses faith in the body as a move-ment of reflexivity through which we can generate *ideals* of ourselves. As we practice and perform these kinetic images, our bodies change. We awaken sources of energy, sensation, and strength within ourselves and *become* persons capable of enacting this reflexive, kinetic-ideal-generating capacity of bodily being. Thus, in practicing contraction and release a dancer *becomes who he is*: he reveals a human as the rhythm of his own bodily becoming, a "doing 'I'." In these respects the *practice* of contraction and release enacts a magic transfor-mation: it provides those who do it with an alternative physical consciousness from which to engage, critique, and revalue western Christian attitudes toward bodily being; an alternative physical consciousness which is in itself a new value—a faith in dance as *theopraxis*. Read in this way Graham's distillation of contraction and release is her answer to Nietzsche's call for a physical–spiritual discipline capable of countering the techniques, innocent or not, that sustain the life of ascetic ideals in us.

A closer look at how Graham stylizes and mobilizes the movements of breathing as the basis for her dance practice sets the stage for understanding how this practice revalues Christian values, and overcomes the physiological contradiction members of Christian culture suffer between a love for dance and internalized prohibitions against it. As in the case of Duncan, Graham's religion language provides the key.

Breathing Matters

My technique is based on breathing. I have based everything that I have done on the pulsation of life, which is, to me, the pulsation of breath. Every time you breathe life in or expel it, it is a release or a contraction. It is that basic to the body. You are born with these two movements and you keep both until you die. But you begin to use them consciously so that they are beneficial to the dance dramatically. You must animate that energy within yourself. Energy is the thing that sustains the world and the universe. It animates the world and everything in it . . . It can be Buddha, it can be anything, it can be everything. It begins with breath.

—B 46[1]

Many explanations have been offered for why Graham chose to cultivate the movements of breathing as an impulse for making movement. Her choice may date to classes she took from Ruth St. Denis in meditation, breathing, and "inspiration," often conducted in the lotus position.[2] Yet at Denishawn, the attention to breathing was not developed into the basis for dance training; St. Denis left technique classes to Ted Shawn who taught a modified form of ballet barre, with some exercises in a turned-out position and some "Egyptian-style," feet parallel to one another (Shelton 1981:127). While Graham also learned yoga at Denishawn, and would have heard members of the Vivekananda Society speak (Shelton 1981:137), she did not seek in the rhythms of breathing a path to the spiritual transcendence of the body, but an entryway for appreciating and cultivating the creativity of bodily becoming. Alternately, Graham may have been forced to develop her own approach to dance training because Ted Shawn threatened to charge her $500 for the rights to use the Denishawn curriculum. Yet is it not clear whether Graham would have been content to rely on that curriculum. While theater artists such as Antonin Artaud were experimenting around the same time with the use of breathing as a way to access states of human experience for dramatic purposes, Graham was the only dancer engaged in this work.[3]

Tracing Graham's religion language suggests an alternative to these interpretations. Graham's focus on the rhythms of breathing fit the criteria she needed to respond to what she had learned from Duncan's attempt to realize Nietzsche's vision for dance as a catalyst in the revaluation of religion. Graham needed a way to communicate through bodily means—and not verbal forms—the inherent creativity of bodily becoming that Duncan's revalued notion of soul represented. In the contraction and release, Graham found a way: in practicing these movements dancers could *enact* their ability to create and become kinetic images of their own life-enabling movement.

As early as 1927 Graham used the terms "contraction and release" to mark "a whole new approach to the physicality of movement dependent upon the breath, and the anatomical changes in the body due to the breathing process" (Horosko 1991:37). In classes during this time, dancers repeated the movements of inhalation and exhalation over and again; Graham watched and commented. She guided her dancers to focus their consciousness on the shape, rhythm, and muscular texture of the actions. As Gertrude Shurr explains: "Martha wanted to know how the body responded when the breath was exhaled; what happened to the bones, the skeletal part of the body. What happened to the muscles, and what was the quality of movement as a result of this activity" (Horosko 1991:38). In noting the changes that occurred in the body in order to breathe, Shurr describes what they discovered: "We found that upon the exhaling of breath, the skeleton or bones of the body moved: the pelvic bone tipped forward, the cartilage of the spine allowed the spine to stretch and curve backward, and the shoulders moved forward, always retaining the alignment of shoulder over hip, while never lowering the level of the seated position" (Horosko 1991:38). Graham encouraged her dancers to isolate these movements and practice them. She discovered the contraction and release.

Briefly, in the contraction, as when exhaling, the hip, spine, and shoulders move in relation to one another as the front muscles shorten and the back lengthens. The lower abdomen curves inward and back, pressing to infinity. In the release, as when inhaling, the movements reverse. A dancer extends out of her lower abdomen and up through a straightened spine. Once a dancer can sense these breathing movements happening in her body and repeat them at will, she can play with them—altering the tempo, direction, and level, or applying basic rules of form by repeating, inverting, or reversing the movement. A dancer, in other words, can discover, in the "natural" movements of her own life-enabling breathing, patterns of potential movement—kinetic images that she can exercise and repeat as a way to enhance her own inner muscular sensitivity and energize her physical consciousness. As Dorothy Bird Villiard describes of this era, the goal was to, "animate the five senses . . . when they are animated you are alive . . . explosive inside the skin—bursting with potential energy."[4] In sum, Graham's discovery of contraction and release provided her with a ground for unfolding an approach to dance practice, making, and performing that embodied a different relationship to the body than that represented by existing techniques of training in dance or religion, one in which the "body" is not a thing but itself a movement to value as a source of knowledge: "My technique is basic. I have never called it the Martha Graham technique. Never. It is a way of doing things differently than anyone else. It is a certain use of the body. It is a freedom of the body and a love of the body" (B 240–1).[5]

In discovering this use, freedom, and "love of the body," Graham provided herself with a ground from which to revalue the concept of *technique* as essential to the *art* of dance in ways that addressed the concerns of both Duncan and Duncan's critics. As in the previous quotation, she did not start out calling her "way of doing things" a "technique"; she reclaimed the word only after developing her "way." On the one hand, in definitions such as

"The ideal of technique is the absence of strain; it is the rhythmic building of the body to a perfect form" (Armitage 1937:102); or "Technique is a language that makes strain impossible . . . An absolute science" (B 249), Graham avoids the criticisms Duncan incurred when describing her dance as movement flowing from an awakened soul. As Graham counters, dance is a discipline—an absolute *science*. Yet in support of Duncan, Graham qualifies her claim: the kind of discipline makes all the difference. For Graham a technique that expresses a love of the body is one that teaches a person to listen to the rhythmic forms of her own bodily movement: "the relationship of the body to itself upon which a dancer's craft is built is an attitude of listening" (G52). As a kind of listening, technique builds the body in line with its own potential for movement, strength, and health. And a technique based on the movements of *breathing* does so with doubled intensity. For not only is a dancer listening to her body, she is listening to the movements that enliven her body. In so far as the movements of breathing are the movements by which a body lives and becomes who it is, the conscious repetition of these movements guides a "rhythmic building" of the body in which the standard for "perfect form" is given by an individual body's own living form. As one of her dancers confirms: "Because these actions come from the depth of the body as the source, I believe her technique never goes against the laws of natural movement in the body" (Yuriko in Horosko 1991:112).[6] In short, Graham defines technique in a way that refuses to impose movement forms upon a body while still providing criteria for a system of dance training. She does so by cultivating the rhythmic movements of a body's living as the source of movement, rather than an awakened "soul" or physical consciousness per se. As such, what is *perfected* in Graham's vision for technique *is* an "absence of strain"—not a set of codified movements but a capacity to make and use kinetic images of one's own breathing as the impetus for dancing. What is perfected is an inner responsiveness of body and mind such that a dancer's desire to dance finds expression in movements that honor and enable the health of her body.[7]

In revaluing technique Graham laid a ground for evaluating all systems of (bodily training in) dance or religion. For example, she admits that there are movements in her technique that also appear in classical ballet.[8] The difference between them lies in *how* a dancer initiates and completes a given movement. A Graham dancer executes a plié by animating the kinetic images of contraction and release. He does not forcefully rotate his knees and ankles into a turned out position. He uses his breath to guide his action and awareness. He learns how to make any movement in a way that expresses a life-enhancing love for his bodily becoming. To practice Graham's technique, then, is to learn how to connect every movement to a source that gives it meaning as a revelation of a *human*. Graham confirms: "Throughout the performance of these technical exercises, a woman remains a woman, and a man a man, because power means to become what one *is*, to the highest degree of realization" (G41:182). Graham's point here is not to reify a distinction between "woman" and "man" but to affirm that the process of becoming one or the other is a function of practicing the kind of bodily movement she teaches. Graham intends her exercises to enable a person to grow and unfold the wisdom of his or her bodily becoming.

Revaluing Religion Language

While scholars tend to overlook the uncanny regularity with which Graham uses religion language to describe the process and rewards of practicing her technique, this language discloses how her work builds on Duncan and Nietzsche's to advance the project they share of revaluing values. Graham describes how practice expresses and generates "faith"; how the fruits of practice include "grace" and "innocence"; how persons who practice become "athletes of god," "divine normals" even "gods and goddesses"; and how practice-enabled dance is an "ecstatic offering to life," a "graph of the heart," revealing the "inner landscape of the soul," and a technique for the "revelation of experience."[9] Such references cannot be dismissed as mere rhetorical devices. Graham does use them as metaphors, but she does so in such a way that she claims for the practice of dancing a place alongside reading and writing as an activity through which the verbal metaphors that govern our sense of ultimate reality secure their meaning. Graham, like Duncan then, frames the practice of dance as a source for renewing "religion." Yet she goes farther than Duncan in disclosing a necessary interdependence between the practices of dancing and writing in relation to religion. Graham's use of religion language suggests that the ability of words to convey religious meaning depends on conditions the words themselves neither provide nor guarantee: the kinds of conscious participation in bodily becoming exercised in the practice of dance. What must be renewed about "religion" is not just its values, but also the forms of its practice that engender those values.

An example of how Graham revalues religion language in relation to dance practice occurs in the preface to a book of photographs of her taken by Barbara Morgan. Graham writes:

> Every true dancer has a peculiar arrest of movement, an intensity of attention which animates his whole being. It may be called Spirit, or Dramatic

Intensity, or Imagination, any word that explains why he does what he does. There is a sweeping line of intent that *services* his entire body. It is very like the act of listening. There is a complete focus upon a given instant.

I know of no other word for this dynamic except *co-ordination*. To me co-ordination means dominion of Spirit-of-body over all parts of the body, until it produces the unity that is passion. It is the activity produced by this Spirit-of-body that is Dancing. It is the organization of this activity that is the art that is Dance. (Morgan 1941:Preface)

Graham admits, this "peculiar arrest of movement" that characterizes a dancer who practices *may* be called "Spirit." Her "may" suggests that Spirit is a word, only a word, and one among many. In so far as it may be used to describe this peculiar intensity, it is the practice of dancing that provides the content for what "Spirit" means. The value of the word "Spirit" here is that it serves to "explain why" a dancer "does what he does." What he does remains constant.

Nevertheless, Graham then notes that the only word she knows to describe this dynamic is "co-ordination." At first it appears that she is moving away from a "spiritual" register to focus on what is "real," translating abstract language into the practical matters of dancing. Yet Graham immediately moves again to redefine coordination as an activity produced by "Spirit-of-body." In doing so Graham moves back and forth between words drawn from religion and descriptions of dance practice such that the meaning of each is pushed and pulled into focus by its tension with the other. Through this movement, Graham revalues "Spirit" as "Spirit-of-body," and later in the passage she moves even further in her revaluation, referring to "body-spirit." The effect of Graham's movement, then, is to communicate that the meaning of the word "Spirit" is something realized in each individual in and through the *practice* of dancing—of listening "in space and within" (Horosko 1991:39) and learning to make kinetic images of one's breathing.[10] This same logic may be applied to Graham's description of dancers as athletes of God, divine normals, or gods and goddesses: in every case Graham lifts such words from any explicitly religious context, treats them as metaphors, and identifies the practice of dancing as defining their content.

In short, Graham uses religion language in ways that claim for the practice of her technique something that Christian religion, at least as manifest in the mainstream modern West, has restricted to verbal languages: a constitutive role in generating the meaning of "religion" (and its words). The practice of her technique develops dance as a "language" not by imitating words but by exercising a reflexivity—and thus a capacity for representation—that is internal to a body's own movement. It is a reflexivity that engages the metaphor-making occurring at the level of sensory experience that

Nietzsche described and Duncan exercised. The realm of sensation and knowledge opened by such practice is not verbal and not nonverbal; it is rather the realm in which experiences give rise to verbal signs and verbal signs prove (or not) their value in enabling physiological health. In short, in her use of religion language to describe the effect of practicing contraction and release, Graham locates a power of (self) representation in an inherent creativity of bodily becoming. To become an "athlete of god" is to exercise this power in becoming who one is; it is to express faith in dance as theopraxis.

How the practice of Graham's dance technique effects this transformation and allows a person to overcome the physiological contradiction of being for and against dancing appears more clearly in a closer analysis of what the practice of contraction and release entails. Practice draws into physical consciousness kinetic images of breathing, center, and spine, and through this process transforms a dancer's experience of himself as body. He comes to know himself differently, as a *dancer*. Not only do these kinetic images give the Graham dancer's movement its characteristic percussive dynamism, they enact, as Graham's religion language implies, a revaluation of Christian values.

Contraction and Release

statement—affirmation
clear, brilliant, deep breathing—
long stretches of the body in space

—N 42

If a technique is built on the movements of breathing, then one who practices it must first of all breathe. Second, she must learn to attune to the rhythms of her own breathing in a manner "very like the act of listening" (Morgan 1941:Preface). Such listening is not performed with the ears. It requires that a dancer become conscious of bodily sensations; it requires that she develop an ability to "hear" subtle pulses of energy passing into movements. Breathing itself invites such awareness. When we breathe, we feel; we open ourselves to our senses. By attending to her breathing, then, a dancer practices extending her consciousness along the length and breadth of her individual body, and disciplining that consciousness to her body's shapes and contours. She learns to empty her consciousness of distractions, abstractions, and ideals, and gather her sensibilities into the present moment. Through this process she can discover the basic movements of spine and muscles described earlier that constitute the rhythmic pulse of the breath, the pulsation of life.

Yet the movements a dancer practices as contraction and release are not identical to the movements of breathing. They are *stylized* images of breathing movement. They are *kinetic images*. Ordinarily small movements are accentuated, elaborated, and in the course of a Graham class, the rhythm, direction, and contour of the movements are varied from exercise to exercise. As a person practices making the movements of contraction and release, he improves his ability to perform the movements quickly, without strain, with greater degrees of force and nuance in timing. He learns to breathe more deeply and fully, sending more oxygen to his muscles and ligaments and enabling them to contract with greater speed and power. In the process, not only does he grow strong, he strengthens the muscles involved in breathing beyond what is required to breathe. This additional range of breathing movement pulls new ranges of sensation into consciousness. He *builds* his body, *rhythmically*, moving farther into a stretch, faster into a spin, higher into a leap, deeper into a contraction. In short he learns how to energize his sensation of body in-movement, every cell alert. As Graham dancer Helen McGehee notes, "dancing should be a source of commanding a deep, inner energy . . . Technique is the means by which you rekindle this vitality" (Horosko 1991:83).

As the patterns of muscular movement associated with breathing come more clearly into physical consciousness as kinetic images and more firmly within the sphere of the dancer's control, they begin to carry more and more of the work required to execute other movements. A dancer acquires the ability to make *any* movement with greater sensory presence. The kinetic images of breathing her practice generates become her "natural" response when prompted to move, regardless of whether the cues come from music, intention, or watching another person move, or whether she is reaching for a box of cereal or brushing her teeth. This ability to mobilize bodily sensation into form is what enables a dancer not only to desire that her foot be pointed, but actually to do so.

In transforming what a body can do, the practice of contraction and release thus transforms a person's sense of what his body is. In so far as what he practices are his own *pulsations of life*, the practice provides him with an experience of his own reflexivity—his freedom—his ability to make kinetic images of himself, and through that act of making, to become capable of generating new movement values. In other words, the practice of contraction and release, though itself an exercise of hard physical work, has ramifications that extend beyond the physical for the ways in which it exercises the inherent creativity and reflexivity of bodily becoming. In developing the kinetic images of contraction and release *I come to know myself as the active, self-generating medium in and through which I become who I am.* As Graham summarizes: "You are in competition with one person only and that is the individual you know you can become. And that is the thing—that is what makes the life of a dancer the life of a realist and gives it some of its hazards and some of its wonder. It is a creative process. It is out of that handling of the material of the self that you are able to hold the stage in the full maturity and power which that magical place demands" (G58, 5).

In providing a person with practical knowledge of her own reflexivity, the practice of contraction and release also provides a dancer with resources for understanding the human experience of others, including her past (images of her) selves. In so far as breathing is a rhythm that accompanies all moments of human life, there is no pattern of breathing that does not represent the life condition that it enables. Every subtlety of thought and emotion finds expression in the tempo and quality, in a catch, suspension, or flooding expulsion of breath. In practicing contraction and release, then, a dancer is working with the "material" of her self in the sense that she is working the accumulated store of thoughts, feelings, actions—and errors— that her body/self is. Conversely, a person's own experiences provide her with resources for exploring the movement ranges opened by the practice of contraction and release. As May O'Donnell explains: "Going with her into her search taught you a great deal about yourself. You could discover where your base was, that point within yourself where your energy is released into movement and dance" (Horosko 1991:60). The work of a dancer who engages in this practice *is* to learn about herself, to become conscious of how her own desires and attachments, her patterns of muscular habit, inform her ability to move. In this way the practice of contraction and release helps a dancer overcome the obstacles in her own flesh to embracing her bodily becoming as a great reason and to embracing dance as an expression of life.

Moreover, as a person develops this faith in his ability to move, every movement he uses these kinetic images to make *represents* the bodily reflexivity that he cultivated to generate the kinetic images in the first place. He reveals himself in his movement as a bodily becoming. His movement reveals the *human* as a bodily process by which humans make and become who they themselves are. As Villiard explains:

> [As] the source and impetus for all movement . . . laughing . . . sobbing . . . shouting . . . whipping, spitting . . . the correct movement was contraction, nothing else. [The movements were] abstracted . . . so the audience didn't know what it was and wouldn't be told, but it would be there . . . Eliminate most of it (else it comes across as the me of the performer) so you can thrust it onto them and they feel it.[11]

A dancer who practices using the contraction and release as the source of his movements does not express his "self"; he expresses his faith in (his own) bodily becoming. What he enables the audience to "feel" as the "sensation of living," the sense of potential and power, is this process of kinetic image making—the *life*—that Graham intends her dancing to affirm.

In this way, the practice of contraction and release that Graham develops meets the criteria for extra-textual work that Nietzsche figures with his images of dance. The act of practicing itself revalues values. While the practice of any physical technique may catalyze a transformation in a person's experience of bodily becoming, practicing kinetic images derived from breathing does so in a way that counters the effects of the ascetic ideal as Nietzsche describes them. It enables a person to know and have faith in her body as a great reason, a "doing 'I.' " It educates her to attune to her senses, listen to her instincts, and thereby stir the sources of energy within her. It awakens her desire to become the human she is. As Graham holds, "Desire is a lovely thing, and that is where the dance comes from, from desire" (G91, 5).

This account of contraction and release as actively participating in the revaluation of values further illuminates how Graham uses religion language in ways that elevate the practice of her technique as a source of religion language's meaning. Where practice generates *faith*—it is faith in one's body as a process of its own kinetic-image making. Where technique serves as a *revelation* of experience, it does so by providing a visual, visceral experience of humans creating kinetic images of the rhythms—the pulsations of breath and life—that inform and enable all human experience. Where dancers are *athletes of God*, they become so as they exercise their freedom in dance in becoming who they are. In a rare autobiographical moment in her *Notebooks* Graham wrote: "Strength comes from a use of the muscles—The athletes of God wrestled & grew strong. They chose, & acted [from Ephesians]. What I do must be done in full sunlight of awareness—I must learn a means of changing the mood—the state of consciousness—without double-mindedness or tho't of gain" (N 87).[12] Graham's claim here is not only that dancing is a practice through which a person can become an athlete of God, but that dancing is a kind of practice that makes the value of practice evident. It is Graham's intent to change the "state of consciousness" that leads people to ignore the role of bodily practice in the process of becoming human, a process that includes generating and believing in (religious) ideals.

Finally, in so far as Graham perceives religion as a "matter of personal choice," her decision to use religion language to describe and teach the contraction and release represents her belief that the "material of the self" people should animate in their practice must include whatever commitments they have for (or against) "religion." Graham uses religion language to reach dancers at those places where they make existential choices about who they are and what they honor. She uses religion language in ways that demand that dancers find the meaning of whatever they value in their practice of technique (as she defines it). She thus invites people to enter a "dancer's world"—a world in which their highest values, their sense of reality, truth, and beauty, their connections to family, tradition, culture, and self, are held accountable to their evolving practice of contraction and release. In so far as the meaning of our religion language expresses our physiological state (as Graham learned from Nietzsche and Duncan), then there is reason to embrace a body's own logic of becoming, when honored, disciplined, and developed, as a medium alongside writing for generating and becoming our highest ideals. The "clear, deep, brilliant breathing" of the contraction and release thus catalyzes an *affirmation of life*.

Center

The first principle taught is body center.

—Rogers 1941:185–6

This center of gravity induces the co-ordination that is body-spirit.

—Morgan 1941:Preface

Always return to center. So be it.

—Horosko 1991:146

The practice of Graham's technique participates in a project of revaluing values by pulling into physical consciousness two other kinetic images as well: center and spine. Where Nietzsche describes how modern persons suffering from decadence lose their center of gravity, Graham designs a technique that educates people's senses to a center, that is, a source and telos of bodily movement. And again, the religion language Graham uses to describe and teach this sense of center traces an interdependent relationship between writing and dancing as complementary if contested activities for generating ideas of and about religion.

In the Graham world, center is a function of movement. It refers to the physiological "place" where a contraction erupts into a release, and a release empties into a contraction. As a dancer contracts, abdominal muscles press back against the spine, hollowing out the lower abdomen. A dancer pulls on the contraction, and thereby, opens a sense of suspension and stillness in an area of the body where the last micro-movements of exhalation occur, usually several inches below the belly button. While functioning as a nexus of life-supporting systems—digestive, circulatory, nervous, and reproductive—this physiological site is also the muscular matrix that a person must animate in order to stand upright, walk forward, and move his limbs freely in relation to his torso. In his classic description of the human anatomy Straus explains: "The center of gravity is elevated high over the small base of support . . . [which] provided for greater flexibility and variability of movement, but it increased instability and the danger of falling. Man has to find a hold within himself. Standing and walking, he has to keep himself in suspension."[13] In learning Graham's technique, a person practices the actions by which a person finds such a "hold within himself."

Center, so defined, is not given. It is not a foregone conclusion that a person will have a center, or be able to find a center. Nor can she create center by a force of mental will. The center is *unteachable*. As one Graham dancer explained: "Graham's desire is to get at the unteachable . . . Her search is for the beautiful, the unteachable, but nevertheless discoverable center" (Helen McGehee in Horosko 1991:86). All a dancer can do is actively invite a sense of center to appear in her physical consciousness. Listen as Graham exhorts:

> Pull, pull on the contraction. Do not cave in. And the contraction is not a position. It is a movement into something. It is like a pebble thrown into the water, which makes rippling circles when it hits the water. The contraction moves. (B 250–1)

As the words suggest, a dancer cannot force herself into the right position for center because there is no such position. A contraction moves, and by moving into "something" it creates a sense of the place from which its rippling circles radiate: "The motivation, the cause of the movement, establishes a center of gravity" (Morgan 1941:Preface). What a person senses as center, in other words, is a function of the movement she develops the ability to make.

This understanding of center further confirms the point made in the earlier section that practicing the Graham technique involves more than hard physical work. In order to animate a sense of center, a dancer must be willing to surrender to the movements of contraction and release—to give his thoughts, emotions, and sensations to those movements and greet whatever obstacles arise in mind or body to their execution. As Denise Vale notes:

> A Graham dancer has to be able to cope physically with a set series of exercises, and the body has to be strong enough to reproduce them. But it goes

beyond even that strength. Martha's dancing isn't about limbs; it starts from the center, and people fight the fact that they have to go to their centers to do it. (Mazo 1991:45)

Going to one's center in this sense means taking a risk. It means feeling one's desire for what one does not yet have. It means opening oneself to the possibility of failure, of falling off the rope. It not surprising, then, that Graham dancer Christine Dakin likens the process to a "ritual which leads us to deepen our awareness, concentrate our energies, expand our ability": a dancer must not only desire to move, but he must have a "willingness to push the limits we believe we have, to enter into each class as a new challenge . . . on the deepest level, it is a life's work" (Horosko 1991:131). It is a "life's work" because it is the work of a life; the "object" of one's deepening awareness is living, always changing. Every day what a person must do to go to his center is different, depending on how he has slept, eaten, thought, felt, and danced. Going to one's center demands an honesty with oneself that must be recreated every day.

Nevertheless, though unteachable, a sense of center is discoverable. A sense of center can emerge, and as it does a dancer is able to use the muscular action of contraction and release to greater effect in powering her weight and aligning her limbs through space. She can summon more force into her opened sense of center and direct it outward with increasing precision. Describing this development Mille describes how "the arms and legs moved as a result of this spasm of percussive force, like a cough, much as the thong of a whip moves because of the crack of the handle. The force of the movement passes from the pelvis and diaphragm to the extremities, neck and head."[14] The trajectories of the dancer's movements become more direct, commanding a greater kinetic, visual, and visceral impact. As the energy erupts outward, it circles back to center, pressing deeper within the body. A dancer thus pumps into circulation currents of energy that she uses to send movement forth in larger arcs of expression. Graham never wanted her movement to "leak out" (Lloyd 1949:50).

In light of this description, the significance of the religion language Graham uses to describe center appears again. When she describes center as the source and "so be it" or "amen" of dance, she confirms, in line with Nietzsche, that the meaning religious ideals have for us is a function of how we participate (or not) in our own bodily becoming. In so far as a dancer learns through contraction and release to experience center, then that experience informs what the word "center" means: center is invited, not made; desired, not forced; open, not closed; empty, not full. Center is a value that a body creates for itself; and Graham, like Nietzsche, insists that human beings should build their centers in their bodies, on earth, and not in an ethereal "soul," a rational ego, or an ideal afterworld. Moreover, "center," defined in this way, is not singular. As Graham confirms: "The body is made so beautifully. There is a little center of gravity in every part of you" (G52). As the practice of contraction and release pumps energy and awareness through the expanse of physical consciousness, every moment of bodily sensation has the potential to come to life as a resource for creative expression. The implication of this revaluation, then, is that those who desire a "center" for themselves need to cultivate a faith in the body, a love of the body, as the practice of Graham's technique is designed to do. Conversely, in so far as dancers use religion language in inspiring themselves to make the movements through which a sense of center emerges in their physical consciousness, their movements will represent this understanding and use of religion language as sourced in practice. The movements themselves will enact a dynamic interplay in which religion language stimulates dancing, even as the dancing revalues the meaning of the religion language used to describe it.

In contemporary literature and art, of course, "center" is a contested term. As a hallmark of the modern consciousness, the "center" is that which has not held and cannot hold: we are rootless, alienated, erring. The institutions that now infiltrate our lives are decentralized webs and nets operating via concealed hierarchies of power; the grand god narratives that once granted coherence in time and space to our sense of reality are tattered; the fabric of community is unraveling. In such a postmodern world our unflagging desire for center is nostalgic, misguided, and even oppressive. Yet Graham's art reveals such perspectives as signs of a Nietzschean decadence. Such perspectives represent the thoughts of bodies cut off from their own creative capacity. From the perspective of her practice, to deny the possibility of center is to shirk one's responsibility as a bodily creator of value. It is to dwell in an illusion of self as a rational thinker. It is to fail to appreciate that we learn by practice and that practice provides the context within which our values and ideals have meaning. It is to seek to reduce our physiological contradiction of being for and against ascetic ideals to a simple surrender to their inevitability. We are lost.

By contrast Graham's pursuit of center through dance, when read through a Nietzschean lens, represents her strategy for overcoming the physiological contradiction of both longing for a firm and trustworthy center and doubting its possibility. Graham is able to experience the lack of center as an occasion to exercise her bodily becoming in recreating a physical sense of center. Her practice thus represents a response to the death of God that Nietzsche describes—a practice in which people can transform their experience of that death from despair to hope, from end to beginning. Through the practice of contraction and release Graham overcomes ascetic ideals by changing their meaning. In the context of her practice an ascetic ideal registers as an opportunity to awaken the span of physical consciousness the ascetic ideal denies. As a freedom and love of the body, Graham's technique incarnates life-affirming values. It guides dancers to reclaim their open-ended value-generating power—their potency as kinetic image makers—*as* their *center*.

Spine

[T]he spine is your body's tree of life. And through it a dancer communicates; his body says what words cannot, and if he is pure and open, he can make of his body a tragical instrument.

—B 8

In relation to a third kinetic image pulled into physical consciousness by (and as enabling) the contraction and release, Graham's use of religion language again provides evidence of how her dance practice participates in a Duncan-informed Nietzschean revaluation of Christian values. Where "center" revalues the source and telos of an individual life from its mental and spiritual ideas or a particular physical location (like the solar plexus) to its bodily becoming, "spine" represents the transformed sense of self in relation to world that practicing contraction and release invites.

Sitting on the floor with the soles of the feet together not only helps a dancer execute the contraction and release in ways that awaken a sense of center as a generative space of turn and return. It also helps a dancer discern the movements of his vertical axis as his spine pushes down against the ground and up against the downward pull of gravity. As the spine curves inward on the contraction and lengthens in the release, a dancer releases the energy of center along a vertical alignment between hip and shoulder. He prepares to stand. For Graham, "the floor was the earth . . . the earth's gravitational pull was not something to be opposed . . . but something to be exploited" (Helpern 1994:25). In pulling on the contraction and release, a dancer uses this tension between gravity and ground to define his relation to the vertical and horizontal spaces of human living.

In likening the spine to a "tree of life" then, Graham invokes a nexus of religion images that this aspect of her practice serves to revalue. As a tree, the spine is a *conduit of life*, rooted in the earth and channeling energy through trunk and limbs. On the one hand, the spine's column connects and separates realms above and below: "There is only one law of posture I have been able to discover—the perpendicular line connecting heaven and earth" (G41, 182). On the other, the branches of the tree/spine reach out horizontally, defining a relationship to others in the world. A dancer's spine is thus a link between the heaven of highest ideals and the earthly roots of desire, an "axis mundi" of sorts, establishing an order of things suspended between high and low, a space of the sacred in the midst of chaos.[15] Her sense of spine defines her orientation, allowing her to move in relationship to what appears to her. A sense of spine as this connecting line increases her ability to stand where and how she chooses. In so far as practice invites physical consciousness of the spine as a tree of life, a dancer comes to sense the "sacred" as a function of the work she does in finding a hold within herself from which to move.

Further qualifying this image of spine as a tree of life, Graham also teaches that, "life flows along a spiral path" (Horosko 1991:116). Before the discovery of the DNA double helix, Graham encouraged her students to watch the spiral paths of a plant toward the sun, of the muscles around the bones, of the energy around the core of the spine. The spiral begins with the contraction of the hip backward, or its release forward, and continues up the body, creating a stretch that pulls in opposite directions around the spine. The whirl of the spiral, then, serves to organize the rest of the body around the line of posture defined by a dancer's sense of center, gravity, and ground. Pulling the body into spiral positions around a central axis, using the contraction and release as the guiding impulses, a dancer strengthens the contraction and release as the source of his power. As Graham confirms, the dynamic of the spiral belies the illusion of stasis conveyed by the tree image. The spine that stands is charged: "I use the word 'posture' to mean *that instant of seeming stillness when the body is poised for most intense, most subtle action, the body at its moment of greatest potential efficiency*" (Rogers 1941:181). Introduced to her technique in the 1940s, the spiral around the spine quickly became a primary means for directing the force accumulated in contraction and release.[16]

What results for one who cultivates this sense of spine as a spiraling tree of life is what Graham calls "grace," revaluing the term as she does.

> The most important thing for you as a dancer is your posture. Grace in dancers is not just a decorative thing. Grace is your relationship to the world, your attitude to the people with whom and for whom you are dancing. Grace means your relationship to the stage and the space around you—the beauty your freedom, your discipline, your concentration and your complete awareness have brought you. (Armitage 1937:101–2)

Traditionally, Christians use grace to refer to God's unmediated action in the world. Grace is a gift, given not earned. In calling upon dance to provide the content for this word, Graham implies that the practice of contraction and release yields such a gift. Her point is not to discount the experience of grace, nor relocate its source from an omnipotent God to a human will. In her use, grace remains a gift, a miracle. However, it is a gift that comes (or not) in response to the hard physical work of cultivating inner responsiveness of body and mind. In other words, it is the practice of contraction and release that provides a person with a perspective for understanding the sense in which grace is a gift: it is a gift that comes in and through our faith in the ongoing mysteries of bodily becoming, in our willingness to discipline ourselves to the rhythms of that becoming, and in our desire for becoming who we are as self-creating humans. Grace is what flows through us, through our spiraling spines, as we bend, twist, and stretch from the centers generated in us through the practice of contraction and release. In this way Graham's practice again provides a physiological perspective for critiquing and revaluing religious ideals that encourage us to discount our bodily becoming as irrelevant to the purpose and joy of living.

Revaluing Ascetic Ideals

When read in the context of work by Duncan and Nietzsche, it is evident that the significance of Graham's discovery of contraction and release extends beyond the creation of a new dance technique. As Nietzsche taught and Duncan confirmed, ascetic ideals cannot be refuted. They are us. So too is the resistance to them that drives us to embrace dance as representing an alternative mode of valuation. Graham does not deny this physiological contradiction; she experienced it herself. Rather, she created a technique of dance that provided her with a physical consciousness—a perspective— from which to appreciate that contradiction as a stimulus to realizing the potential of dance as theopraxis: as a medium for generating ideas of and about religion. Through the practice of her technique, people may come to experience their own latent hostility toward their bodies as a resource for both expanding their own movement potential, and developing forms of dance capable of expressing a love of the body. In so far as the listening required to attend to the rhythms of breathing leads them to discern the workings of the ascetic ideal in their relationship to themselves, they know why the practice of dancing is necessary. Through practice, we learn that practicing a technique akin to Graham's teaches us to exercise our own kinetic image making potential in becoming who we are; it is what enables us to transform our experience of physiological contradiction into a reason to love life. We become moments of culture in which Christian values are overcoming themselves.

More broadly, the practice of contraction and release provides resources for rethinking dualistic structures of logic endemic to western philosophy and theology and not just values pertaining to the body and dance. The logic of contraction and release is not strictly speaking oppositional; nor does the practice of the two movements aim at their unity. Rather, the practice of Graham's technique embraces the tension between contraction and release as itself generative of sensation and experience. Graham's technique is designed to enhance this tension in order to open a sense of breath, a space of center, and a span of spine. It teaches a dancer how to sustain within herself the contested interplay of those forces and appreciate that interplay as the source of her virtue, her values, her dance. It is by amplifying the difference between two interdependent flows of energy that a dancer builds her body in line with its own form, expressing her faith in and love for her body.[17]

In sum, by recreating the movements of breathing into the kinetic images of contraction and release, Graham not only exercises the reflexivity that impels bodily becoming, she does so in ways that reveal a body as a symbol of itself—as always already in tension with the images of itself that its movement is constantly generating. Graham animates and accentuates the rhythmic turns where mind becomes body and body mind, where will flows in to action and action generates a sense of self, where a person is and participates in his own creation, where he creates his body/self. Graham develops her body into what Nietzsche called a "very subtle and reliable instrument" (EH 241). In her words: "Perhaps what we have always called intuition is merely a nervous system organized by training to perceive" (G41, 180).

In these respects, the training Graham designs her practice to offer is one that educates our senses in contrary and complementary directions to those taught by practices of reading and writing and enforced by ascetic ideals.[18] The practice of contraction and release occurs in a space between pure experience and verbal language—a realm of primary metaphor-making—in which people can cultivate a knowledge that eludes verbal representation. This knowledge is not knowledge of how to execute a pirouette but rather knowledge of how to energize oneself in the process of one's own bodily becoming—or, as Graham so often said, how to be "reborn to the instant." As she intones:

> In order to work, in order to be excited, in order to simply be, you have to be reborn to the instant. You have to permit yourself to feel, you have to permit yourself to be vulnerable. You may not like what you see, that is not important. You don't always have to judge. But you must be attacked by it, excited by it, and your body must be alive. And you must know how to animate the body; for each it is individual. (B 16)

Graham thus creates a dance practice that guides people in knowing themselves in the ways Nietzsche advocates—in and through the ladder of their own experiences. She does so as well, as chapter 8 explores, in ways that allow her to carry on Duncan's critical advance of Nietzsche's treatment of the role played by "women" in the revaluation of values.

At the same time, the self-knowledge that Graham's practice provides does not refute or replace the knowledge provided by education in ascetic ideals; it provides a complementary perspective on what those ideals represent as true. As Graham figured out how to energize her physical consciousness of contraction and release in making a greater range of movements, she explored new existential states, thoughts, and emotions, finding more ways to abstract, enhance, and articulate the pattern of contraction and release through a variety of positions—sitting, kneeling, and standing. In particular, she used the contraction and release as a medium for interpreting texts of and about religion. She used the physical consciousness and kinetic images her practice opened in her as resources for *reading*. In making dances, she sought to recreate the physiological forms that enable the meaning of texts that continue to haunt the religious imaginations of many in western culture. How she did so is the subject of chapter 8.

Chapter 8

Words to Dance

What are the dances to be? | The supreme universal moments—in self-realization which are common to all lives.

—N 217

I'd like to be known as a storyteller. I have a holy attitude toward books. If I was stranded on a desert island I'd only need two, the dictionary and the Bible. Words are magical and beautiful. They have opened up new worlds to me. I'm a very strange reader; I like espionage. Nothing lets me think more clearly through a problem than reading and alternating between two mysteries at the same time.

—B 274

She is standing at the back of the stage, in the center. She moves forward slowly, thrusting one hand down and out before her; the other hand down and out before her. The voices of two singers meld in dissonant tones: "Clytemnestra. Why dishonored, why among the dead." She is Clytemnestra and the question she poses to herself through her gestures is the question Graham as a choreographer and dancer posed to the text by Aeschylus that inspired her dance: Why Clytemnestra? Why, at the end of the trilogy that comprises the *Oresteia*, does Clytemnestra alone suffer? She teamed up with Aegisthus, her lover, to kill Agamemnon, her husband, who had killed their daughter, Iphigenia, and is killed by her son, Orestes, in return. Amidst this blood bath, however, she is the one left to roam the underworld. Why?

With this dance, first performed at the Adelphi Theater, New York City, on April 1, 1958, Graham, aged 64, enacted a paradox at the heart of her dance making—a paradox shared to greater and lesser extents by many if not most of her dances.[1] On the one hand, as she insisted repeatedly: "The dance is not a literary art and is not given to words—it is something you do" (Armitage 1937:103); or as noted, "My dancing is . . . not an attempt to interpret life in a literary sense" (Armitage 1937:102–3). Yet, on the other hand, from the beginning to the end of her dance making career, Graham mined classic texts from world history—especially texts of and about religion, the Bible, and the dictionary—for word images, story lines, characters, and themes to inspire her kinetic imagination. She read constantly, often reading several books at once. She copied down quotations that interested her into notebooks, only some of which have been published. Nearly all of her dance titles, characters, and themes bear some more or less explicit relation to a textual inspiration. In short, Graham refused to talk about her dances despite a professed "love" for words and a "holy attitude" toward books. She denied that dance is a literary art yet frequently choreographed texts. How are we to make sense of these apparent contradictions?

When read as an attempt to engage and advance Nietzsche's project along the trajectory opened by Duncan, Graham's ambivalence toward books appears as a strategy for realizing dancing alongside reading and writing as a medium for generating ideas of and about religion. Duncan too read avidly for ideas to flesh out and justify her dance vision; and Duncan, as described in chapter 4, argued for the place of dancing alongside verbal expression as a distinct and complementary art. Yet it is Graham whose dances and whose process of dance making enact a conflicted interdependence between dance and text. Her dances present themselves as communicating the meaning of the texts to which they allude, even as the texts appear in the dance as occasions for realizing this revelatory potential of dance. In so far as Graham succeeds in animating this interplay, her dances offer critical perspectives on the textual traditions that comprise the weave of western culture. To the texts she dances, she poses the questions Nietzsche also asked: Does this book *dance*? Does it catalyze in readers an ability to affirm life? Of what value is the value we accord to acts of reading and writing?

In taking on these issues in her dances, Graham stakes a claim whose significance subsequent generations of modern dancers, dance historians, philosophers, and religious studies scholars have yet to appreciate fully: *dance* can reveal what the practice of writing teaches us to ignore, namely the dependence of verbal meaning on our bodily acts of kinetic image making.

After examining the attitudes toward words and texts Graham expresses verbally, this chapter offers a close analysis of her *Clytemnestra* and of the process by which she moved from her reading of Aeschylus's tragedy to the creation of dance movement. I argue that Graham accomplishes in this dance the magic transformation Nietzsche identified with Attic tragedy: her dance provokes us to experience Clytemnestra's tragic tale as communicating participation in an affirmation of life. Graham's dance presents "dance" as a therapeutic theopraxis. Analyses of two other dances, *Errand into the Maze* and *"Acts of Light,"* corroborate this thesis.

Words, Words, Words

It was because Graham loved words so much that she guarded her dances fiercely against them, especially in the early years of her post-Denishawn career. Graham appreciated the power of words, their beauty and magic, their precision and clarity. Recall that, first dance lesson aside (see chapter 6), Graham was socialized as a reader and writer, a Christian and an American, before she began dancing in earnest. Prior to attending Denishawn at the age of 22, she had expressed her artistic interests through writing and acting. As her vision for dance evolved, striking an ideal, enabling relationship between dancing and word-use remained a primary concern.

As a dancer with Denishawn, moreover, Graham had been encouraged to read, mostly texts of and about religion, as inspiration for her dancing. St. Denis and Shawn sought out libraries when they traveled, collecting books on dance traditions of religions from around the world and throughout time, most notably, from the classic traditions of Southeast Asia and the Middle East. During her Denishawn days, the role for which Graham was most famous was the lead in *Xochtil*, a story of an Aztec princess based on an Aztec myth Shawn had discovered. Recall as well that Graham first read Nietzsche, Schopenhauer, and other German philosophers with Horst during this time. As part of her training in dance at Denishawn, then, Graham developed a practice of reading for perspectives on, principles concerning, and ideas for her dancing.

It may have even been a growing sense of disparity between what she was reading and what she was dancing that helped propel Graham to leave Denishawn and the Follies and embark on her own career. St. Denis and Shawn adopted a largely literal interpretation of texts: their dances acted out stories, often using pantomime or gesture language.[2] Impelled by her reading of Nietzsche and by Horst's influence, Graham began to reconsider the role of texts in the process of making dances. As noted in chapter 6, she wanted her dances to register as significant movement, that is of America, and reveal a human. Toward this end, as described in chapter 7, she discovered and developed the contraction and release as the basis for a movement practice. Still, she could have continued making dances as had St. Denis and Shawn by assembling her contraction-and-release-impelled movements like words to narrate the plot of interesting stories. She did not. As is already evident from the analyses of *Heretic, Lamentation,* and *Primitive Mysteries,* Graham resisted the temptation to make dances that functioned as literary events on the model of texts. Instead, she set out to reveal what she later called the "hidden substratum" of the text: the physiological conditions that enable the words of a text to impact hearts and minds. What appears in Graham's dances is a contested interdependence between reading, writing, and dancing that revalues Christian values and realizes a vision of dance as theopraxis.

Not Nonverbal

Many of Graham's protests against perceiving dance as a "literary" art date from the post-Denishawn decade of her career (1927–1937). While scholars tend to interpret Graham's statements as promoting a notion of dance as "nonverbal," a closer look suggests otherwise. What Graham resists are attempts to explain dances in verbal terms as a narrative, or as having a conceptual meaning: "If you can write the story of your dance, it is a literary thing but not dancing" (Armitage 1937:102). As Graham's statement implies, it is *possible* to write a story of a dance, but the dance for which a story can be written is a "literary thing," a pantomime of a drama, and not the "significant movement" she intended her dancing to be. To write the story of a dance is to "rationalize" it: it is to reduce the impact of a dance to a moral, conceptual, or intellectual meaning. As Graham warns: "There is a danger in rationalizing about it too much" (Armitage 1937:103). If we watch a dance with the intention of explaining it, we prevent ourselves from receiving what the dance has to offer. Asking "what does it mean?" we fail to allow our sensitivity to be quickened by the kinetic experience of the dancing (B 202). We express our own impoverished physiological condition. As described in chapter 1, Nietzsche made a similar point about the attitudes toward music he observed in his time.

In resisting the power of words to explain, then, Graham was resisting the logic of the ascetic ideal whose operations Nietzsche exposed in his analyses of desensualization and oversensitization. Graham's point is *not* that dancing is nonverbal; such a perception would reinscribe the logic of the ascetic ideal (mind versus body) that Graham, Nietzsche, and Duncan contest. Graham's point is rather that the intent to read dance as a text expresses a lack of faith in the body and in dance; it expresses an inability or unwillingness of a body to create beyond itself by engaging in the sensory levels of metaphor making. What is important to Graham about a dance is not whether people understand it, but whether they allow themselves to be moved by it (B 202). Thus, when she states that dances communicate "what words cannot . . . a feeling, a sensation" (G85, C1f), she is not claiming that dancing is a bodily thing, operating independent of thinking. Graham's dances are highly intellectual. She is rather claiming that the kind of thinking that responding to a dance demands is what Nietzsche called thinking *through* the senses. In this reading the primary impulse driving Graham's resistance to words is a desire to open a space *between* word and body so as to disclose a conflicted interdependence between writing and dancing at work in the creation of both dances and texts.

For Graham, the conflicted interdependence between dancing and writing goes all the way down. Although words cannot be called upon to explain, narrate, or justify a dance, no dance can appear without the mediating influence of words. Although words may pretend to float free of the bodies who birth them, all words bear traces of their passage. There is no dance without word-use; no word-use without symbolic bodily movement. Even practically speaking, dancers must be literate creatures if only to invent a title, rent a theater, select music, negotiate costumes and lights, rehearse dancers, and coordinate publicity in a verbally oriented culture. What Graham rejects, in other words, is the attitude toward bodily becoming expressed in either the attempt to explain dance in words, or the attempt to isolate dance from words as nonverbal: both approaches to dance reinforce the logic of the ascetic ideal. The challenge for a dancer in modern western culture, in this regard, is to become conscious of the ways in which she cannot *not* speak, read, and write in the process of making of her dances. A dancer must find ways to mark a difference between dancing and verbal practices *without* reinforcing a distinction between verbal and nonverbal that always, by definition, results in privileging the verbal over the "non."

Read through this perspective, statements of Graham's that seem contradictory appear to trace a necessary and generative tension at the heart of her dance making process. When she insists that "the theme of a dance must not be perceived intellectually" and that "the truth must want to be released through body speech" (G50, 22), she is urging dancers and audiences not to allow their verbal education to dominate their sense of what dancing can be or mean; the body has a language too. By contrast, when she avers: "I love words very much. I've always loved to talk, and I've always loved words— the words that rest in your mouth, what words mean and how you taste them and so on. And for me the spoken word can be used almost as a gesture" (G85, C1), she is calling dancers and audiences to remember that the act of using words is itself a bodily matter. Correlatively, even though

Graham strategically blurs any sense of a clear difference between words and gestures, she rejects the idea that either can substitute for the other. Nor do words and bodies simply trace each other's limit. What appears in Graham's words and in her dances is how the process of learning to use each medium is essential for realizing the communicative potential of the other. Where writing exercises the power of the word to describe "the body"; dancing exercises the power of bodily becoming to generate "the word."

Appearing alongside verbal practices in this way, the dance Graham describes serves her as a critical perspective from which to reflect back on the assumptions we hold about the adequacy and authority of verbal communication. Graham employs words and texts in her dances in order to transform our experience of words and texts. Graham dances texts with which we are already familiar, whose endings we already know. She does so to help us become conscious of how bounded our interpretations of these texts are by physiological conditions. Graham's dances engage texts in ways that (re)awaken us to the primary levels of metaphor-making at work in our reading, and thus to our responsibility in reading and writing as creators of value. She reminds us to hold the words we use accountable to the bodily health they express and enable. In sum, Graham engages words *as a dancer*, through the physical consciousness and kinetic image making ability her dance practice has cultivated in her. Just as the practice of contraction and release accentuates and amplifies the difference between contraction and release as a source for heightened nerve sensitivity and dynamic movement, so too does Graham's dance making process animate the difference between dancing and writing as a source for a transforming, healing art that communicates participation in an affirmation of life.

American Document

While commentators often identify *American Document* (1938) as marking a shift in Graham's work, in that she begins to incorporate written materials in her dance, the contested interplay of dance and text enacted in this piece is already evident in earlier work.[3] As described above, Graham was, from the earliest moments of her solo career, mining texts for inspiration (as in *Dance*) and thematic material (*Primitive Mysteries* and *El Penitente*). She was using words—verbs and adverbs but not nouns—to help her explore the expressive potential of the contraction and release.[4] Graham would trace the etymological history of words to their physiological roots, and seek to recreate these physiological experiences in dance. In all of these ways, Graham was already treating words as arrested gestures, or "petrified movement," and seeking to remember the movement they both reveal and conceal.[5]

Even so *American Document* remains significant in relation to the argument just introduced. Here, for the first time Graham fleshes out in dance the implications for a text-based religion and culture of her vision for dance as significant movement. In this piece *dance* appears as integral to the process of interpreting American documents. Dancing appears as the medium through which we can acquire the sensory education and physical consciousness we need in order to be able to understand the texts that comprise our own culture, in all their tragedy and glory, as stimuli to creating life-affirming values. In other words, this dance represents "dance" as having the potential to alter our relationship to words and texts and change our experience of them such that we come to understand documents as arrested gestures, dance as re-membered words, and our bodily selves as rhythms of their own kinetic-image-driven becoming.

There are numerous elements of *American Document* that support this reading. For one, Graham does not minimize the challenge that interpreting American texts entails. She selects texts that represent divergent views, and texts that recount the mistreatment of bodies in American history—the enslavement of Africans, the annihilation of Native Americans, the repression of women. Documents featured in this dance include the Emancipation Proclamation, a speech by Chief Seneca, The Declaration of Independence, sermons by Jonathan Edwards, "The Song of Songs," poetry by Walt Whitman, Lincoln's Gettysburg Address, and others (*American Document*, program, March 10, 1940).[6]

As important as Graham's choice of documents is the perspective on them that her dance affords. The frame of the piece suggests a minstrel show—evoking at once the history of slavery and the use of performance genres to generate critical perspectives on racial and class-based oppression. Within the dance, dancers enter and exit, individually and in groups, indicating the multiple and diverse bodies represented (or not) by this "document."[7] Most significantly, the dance movements that comprise the piece engage and deflect the meaning of the texts whose words provide the accompaniment to the movement.

One illustrative case is the section where Graham dances a romantic duet with her lover (and later her husband), Eric Hawkins. While the couple dances, the soundtrack alternates between passages from the fire and brimstone sermons of the Christian evangelist Jonathan Edwards, and selections from the erotic poetry of the Hebrew Bible's "Song of Songs." In choosing these texts, Graham dramatizes the physiological contradiction characterizing American Christian consciousness—the one that Duncan also wrestled to overcome in her commitment to dancing the "ideal" of "woman." On the one hand, Edwards' sermons exemplify a "Puritan" repudiation of the senses and bodily life as the source of sin; they offer instead a vision of heavenly bliss. On the other hand, the erotic metaphors for human relation to God featured in the "Song of Songs" suggest how steeped in the experience of sensual richness Christian notions of mystical paradise are.[8] Yet against this background of warring words, the lovers' duet proceeds, in spite of or perhaps because of the textual, sexual conflict. As such this piece suggests that "dance" is a medium in which we can identify, name, and affirm the physiological contradiction embodied in Christian texts and in conflicted American attitudes toward sexuality as a stimulus to creating new values. Through the dancing, the audience gains a critical perspective on these texts that highlights the ways in which they have shaped the American experience of being body. In other words, through the experience of watching the dance, we are encouraged to develop a physical consciousness of our own bodily becoming and reclaim our responsibility in evaluating any and all religious values and ideals based on whether they (enable us to) *dance*.[9]

In this passage from the dance and others, then, Graham uses dance to transform our experience of the texts, in this case explicitly Christian ones, that inform American identity. Texts that, from the perspective of one trained to the ascetic ideal, appear to contradict one another, from the perspective of one trained to contraction and release, appear as affirming the capacity of a body to create beyond itself. The paradoxical effect of the dance is that it affirms words and documents as important for American identity in so far as they provide a stimulus for revealing in dance the bodily becoming they presuppose and deny. The movement between dancing and reading and writing appears as the rhythm comprising the *American Document*.

Notebooks

> It happened
>   Once upon a time—
> It happens
>   Every day

—N 83

Further evidence of how Graham used words as catalysts for revealing the potency of dance and revaluing Christian values emerges in her published *Notebooks*.[10] Chronologically, these *Notebooks* represent the middle swath of Graham's long career, roughly 1943–1966.[11] During these 23 years, Graham choreographed some of her most important works, many of which feature direct allusions to classic texts from western religion and literature.

In reading these *Notebooks* two caveats must be kept in mind. First, these *Notebooks* were not written to be read as a book. They were written to remember words, as a repository of images and ideas. They were written to be danced. As a result, it is difficult to determine what a cluster of words or quotations *means*. Graham's entries are not marked by references as to when they were written; it is not clear when one entry ends and another begins, though a few dates do appear—February 18, 1950 or November 19. In addition, while the book's chapters are marked by titles of dances, some of these dances were never choreographed, and the dance titles themselves are not arranged chronologically. Entries for *Dark Meadow of the Soul* (1946) appear after the entries for *Ardent Song* and *Voyage* (1953 and 1954). Nor are the title headings necessarily helpful in identifying the contents of a chapter, or locating notes on a particular dance. One of the chapter headings reads, "The Trysting Tent, Notes contributing to *Ardent Song, Alcestis, I Salute My Love, Canticle for Innocent Comedians, The Triumph of Saint Joan*" (N 215). Meanwhile, "The Trysting Tent" and "I Salute My Love" are two images that pepper Graham's notes for other dances but never appeared as dances themselves; while notes for *Ardent Song, Alcestis, St. Joan,* and *Canticle* all appear in at least one other chapter. "I Salute My Love" is also featured in its own chapter. Amidst this rich conglomeration of words, there is no narrative line and few textual conventions to help mediate the meaning of the text from author to reader. The value of words for Graham—and reader—lies elsewhere than in the form of a book.

A second caveat follows: it is difficult to determine the relationship of the quotations Graham copied down from her reading to what she herself thought about them. Rarely does she write "I." It is impossible to determine whether quotations represent stances she holds or contests. While it is probable that these quotations captivated her imagination in some way, and represent some element of life that she found necessary to express in dance, her choice of quotations nevertheless represents an indissoluble blend of her intent to remember a range of human experience beyond her own and the mobilization of her own experience to that end. Her choices are best construed as imprints left by a dance-trained curiosity impelled to read and write by a desire to discover significant movement.

For these reasons, reading the *Notebooks* demands a custom-made hermeneutic. Such a hermeneutic begins with Graham's patterns of word-use. Even though, as noted, she does not date her entries, Graham does keep track of where she finds quotations—jotting down the author's last name and the page number, and capping the citation with appropriate marks, at least for the first mention. After the initial citing, however, the signs of origin drop away and the word-images or fragments of them literally float on the page in relation to one another. Graham plucks out images from the paragraphs she copies, and copies and recopies the stripped clips into different configurations. She pauses to assess the patterns created, offers a marginal comment in her own voice, and begins again. It is as if Graham is manually moving the word images in relation to one another, bumping them against each other in order to release their potential. She collects word images as one would collect pebbles or shells, fingering their smooth surfaces and finding patterns that highlight the beauty of each. Amidst the transplanted word images, Graham interjects the name of a company member, an idea for a prop or costume or setting. She waits for the images to loosen into movement.

In this use of words, Nancy Wilson Ross, editor of the *Notebooks*, sees what I do: "tentative shapes moving towards realization." Ross writes:

> Is it perhaps not so much a lack of courage as over-concern with the end-products of their gifts that disinclines most artists to leave us more than the vaguest hints of how, in the secret recesses of creativity, tentative shapes moved towards realization? Or is it genuinely beyond their powers to practice Graham's special kind of attentive inattention, at once precise and unfocused and as delicately precarious as that difficult exercise of holding on to the swiftly vanishing images of a dream? (N x)

By focusing on how Graham copies and recopies and re-places her word images in relation to one another, it is possible to recreate, at least in part, a sense of how she reads, and of how she uses texts in making dances. It is thus possible to assess the value that reading and writing have for her in her dance process.

To begin, in leafing through the pages of the *Notebooks* it is evident that the word images that most arrest Graham's kinetic imagination are ones that concern what people in the west recognize as *religion*—ideas about the soul, gods and goddesses, fate and faith, necessity, choice, and freedom, beginnings and ends.[12] Frequently Graham muses over the lives of religious figures (Joan of Arc, Abelard and Heloise) or biblical heroines (Eve, Judith) trying to locate the crux of their motivations and commitments. She is most interested in existential questions: the moments of "self-realization" in which a person makes a decision that determines the course of his or her life;[13] the moments that catalyze a process of becoming who one is and reveal a human being. As she writes: "Great decisions have to be made in life, which, once made are irrevocable, & dominate the man's whole career & conduct afterwards" (N 270). In researching such decisions, Graham's patterns of word-use communicate more than a tumble of images. Her reflections and juxtapositions suggest that she was using these words as images—word images—in order to help her flesh out the role played by words in relation to bodies and by dancing in relation to writing in the act of becoming human. In Graham's patterns of word-use we see her engaging dance as a medium for contesting and revaluing the strands of Christian culture that privilege words over bodies and word-using over dancing.

One of myriad examples appears in the section entitled: "Center of the Hurricane/Permit me to Try Again." This "dance" was never realized, but then again, it is not clear that Graham had (only) a dance in mind. She writes: "And how did all this begin—perhaps to be a book, or at least, something enclosed in covers, with a jacket" (N 301). Perhaps she was thinking of an autobiography, though she seems to have cosmology in mind when she says "all this." Her thought does suggest that someone, if not her, maintains a privileged relationship between books and beginnings, as if books are the place to discover or document where being begins.

Earlier on the page, the chapter opens with a knot of quotations raising issues of writing and remembering. These include: "Drink not too deeply of the River of Lethe," and "One writes to recover a lost innocence" (N 301). The first warns the "soul" against too much forgetting as it passes from life to life; the second celebrates writing as a means to "re-cover" an innocence lost and forgotten in order to begin again. Graham follows these lines with two instances of eyes opening: Christ healing the blind; and Adam and Eve acknowledging their nakedness. The question arises, what kind of forgetting is involved in writing to remember the beginning? Does such writing heal our blindness or precipitate our fall? Does it represent an act of seeing that must constantly be re-written, covered over, forgotten?

Over the next few pages (and elsewhere in the *Notebooks*), Graham repeats her question about beginnings and offers various responses. She cites the Bible ("In the beginning was the Word") and Shakespeare ("To be or not to be"), making a connection between "being" in general and acts of reflection that prompt "rebirth" such as a decision to be, to assume one's heritage, or to "journey towards wholeness" (N 305–6). She finds a link between these two in the creative Word, an "act of statement." Yet she contests the finality of any word and the beginnings it remembers and forgets: "How does it all begin? I suppose it never begins, it just continues—Life: generations / Dancing" (N 302). Dancing here is the living image of Life (as) becoming.

In a later section Graham fleshes out the challenge that this "Dancing" represents for one who looks for beginnings of Being or beings, self or world, in books.

Gesture as language
Language as gesture
gesture as communication—

the visibility of the experience—
Warning— . . .
Gesture is the first (language)
—speech—

Perhaps you argue sound comes first—but movement precedes sound—
& movement is the seed of gesture— . . .
It all goes back to one small boring word— . . .
a clear image of the activity of God. (N 324–6)

For Graham, if "in the beginning" is a word, then the word is not the beginning, but a continuation of the *movement* that makes the sound. From this
perspective, Graham reevaluates: gesture, the seed of dance, and sound, the
seed of verbal language, emerge together as a function of each other. Gesture
is an image, a kinetic image, of a movement that "comes first" in relation to
the word that speaks creation into existence. Note that Graham is not suggesting that the gesture itself *is* the movement before the sound; gesture, like
speech, is a reflection, a "clear image of the activity of God." Rather, the
movement of the sound gives rise to gesture even as the sound of the movement gives rise to words. Pressed to assert priority, Graham holds that gesture
comes first. Gesture is the language on which our use of words is predicated.

The idea-complex Graham introduces here is evident in how she uses
texts in making dances. What Graham intends to discover through her
use of contraction and release and express in dance is a kinetic image of the
movement that precedes and accompanies the sound represented in verbal
images, whether that sound is the one that creates the universe or the one
that exclaims "I am." Graham looks to texts to find accounts of such decisions (or sounds) and uses her trained physical consciousness to recreate
kinetic images of the movement such a step to self-realization entails. The
"body says what words cannot," then, by making the movement required
for self-knowledge *visceral*. Here Graham quotes "Matthiesson-108," on
the power of poetry to "make us from time to time a little more aware of the
deeper unnamed feelings which form the substratum of our being, to which
we rarely penetrate; for our lives are mostly a constant evasion of ourselves, &
an evasion of the visible & sensible world" (N 284). In her words: "It is,
in each instance, less the actual story, but rather the making visible the inner
substance from which the story evolves" (N 326)—where the "inner
substance" of the story *is* the "visible & sensible world" of those living it.

A closer look at *Clytemnestra* (1958) serves to illustrate how Graham
engages texts in making dances so as to transform our experience of them by
communicating participation in the sensation or inner substance of living
they presuppose. Among Graham's dances, *Clytemnestra* is unique. Not only
is this piece considered a pinnacle of Graham's career, and a "landmark in

American theater,"[14] it is Graham's only evening-length work. This dance, moreover, bears a close relationship to the text that inspired it: an ancient Greek tragedy, written by Nietzsche's favorite Attic dramatist, Aeschylus.[15] Graham choreographed the piece and danced the title role.

Within the *Notebooks* themselves, this piece is unique as well. In four sections featuring this dance, Graham's notes include close studies of the text of the *Oresteia* as well as detailed descriptions of actual movements that make up the dance, Clytemnestra's solos in particular. As she insists, "I read everything . . . every commentary"; I steeped myself in Greek myths to find and explore the "image-making power of man."[16] Passages copied into the *Notebooks* from commentaries are broken with squiggly lines in the margin marking her own reflections. She begins with "Perhaps" and tries out different approaches to the story (N 271). In the course of these sections Graham researches the use and origin of tragedy as a genre; collects cross-cultural images of "net"; charts the structure of the *Oresteia*, its seasons, settings, and characters; and continually returns to the beginning of the story, experimenting with ideas and filling them out. Interspersed with these studies, she offers interpretations of character and action, and every once in a while, a piece of her own memory. She distills the text into composite images, and pulls on the resources of her library and experience to illuminate the significance of these images in relation to the events of the story.

Thus, by mapping Graham's patterns of word-use through these sections, it is possible to track her movement from text to dance, from words to bodies, in such a way as to understand how and why the dance that transpires is not an interpretation of the text, nor a literary event, but rather an attempt to provide kinetic images of the movement that precedes the sound—the inner substance of the text. Through her dance Graham offers us an occasion to experience Clytemnestra's horror as a stimulus to love life, and thus to experience dance as theopraxis.

Clytemnestra

Graham calls her *Clytemnestra* "a drama of re-birth" (N 268). It is a dance in which Clytemnestra passes from the "Underworld of the Imagination," her private hell, into a place of affirmation. Like Zarathustra, Clytemnestra gives birth to herself by learning to shatter the tombs of the past locked in her own flesh. She identifies her participation in that past, honors her experience as an occasion for self-knowledge, and in the process, transforms her "it was" into an "I willed it." In the dance, Clytemnestra accomplishes this redemption by remembering. She remembers the events recorded in the

Oresteia three times: she killed her husband, Agamemnon, for sacrificing their daughter, Iphigenia, and is killed by her son, Orestes, in return. "In agony reveal wisdom," the singers plead.[17]

In a 1973 interview when asked about the relationship between her dances and the dramas that inspire them, Graham discusses her intention. She is not interested in "playing out the play," but in communicating the "inanimate thing which motivates" the play. For Graham, a dance that follows the story is a "bore." She assumes that the audience already knows the story which is one of the reasons she chooses it in the first place. In her words: "I'm not interested in that which is seen, but the interior landscape, that which is not seen . . . which everyone has. We have to quicken ourselves to it" (G73). As this paradoxical phrase "interior landscape" suggests, the "not seen" she seeks concerns a dimension of human experience that the practice of dance allows us to discern. As a "landscape" the motivating "not-seen" is a world in which we live and move; as "interior" it is a realm we can only enter by "quickening" our senses, heightening our nerve sensitivity, and knowing ourselves as bodies becoming. In other words, in dancing a play, Graham intends to use her physical consciousness of contraction and release as a resource for viscerally identifying with the figure of Clytemnestra, imitating Clytemnestra's gestures, living her question, and through this process, sensing in herself the physiological conditions expressed in the dramatic action. In this way Graham intends to reveal a power of the text that is not accessible through rational perception alone—its ability to reveal a human, to stimulate the human act of kinetic image making, and catalyze an affirmation of life. She thus reveals *dance* as that practice in and through which we cultivate a life-affirming relationship to texts.

Against the background of these intentions, the genius of this piece appears. The question Clytemnestra poses to her own past in the dance is the question Graham herself poses to the *Oresteia*: why? The dance through which Clytemnestra transforms her experience of her own past is the dance in which Graham, in becoming Clytemnestra, transforms her experience of the text's tragedy into an occasion for her own creative action. In other words, the transformation in which Graham's dance communicates participation is not only Clytemnestra's; it is Graham's. It is the experience Graham wants her audiences to learn to have themselves in relation to a text whose story they already know. Understanding how Clytemnestra makes her transformation is the key, then, to understanding how Graham reads—and wants us to read—the founding texts of western civilization. It is the key to why Graham dances these texts.

In the dance, as noted, Clytemnestra finds her answer by remembering. The dance unfolds as a series of flashbacks in which she remembers her life three times.[18] Single events surface first, pulling with them characters and proximate events, until the shape of a narrative appears. In the first round of remembering, the ghost of Clytemnestra watches while a vision of herself dances. She sees Iphigenia, splayed for sacrifice; she sees the rape of Troy, and Orestes and Electra plotting against her. In the second pass, the ghost of Clytemnestra watches as each character presents his or her self to her, as introduced by the singers. In the third pass, Clytemnestra dances as herself through the return of Agamemnon, his murder by her hand and the revenge of Orestes against her. In each cycle, Clytemnestra pieces together more of the story, and participates more in the retelling. In the end, she experiences "Wisdom Birth Rebirth."[19]

Clues to Clytemnestra's—and Graham's—Wisdom Birth appear in the patterns of word-use in the *Notebooks* that chart Graham's movement from the text of the play to the dance. In the first and shortest of four chapters of notes on *Clytemnestra* written six years before the first performance, Graham sketches an initial reflection that, in hindsight, serves as the seminal idea for the structure and content of the dance. She writes:

> Clytemnestra
>   opens in Hades
>   "Why?"
>   her question
> finishes at beginning before Agamemnon—
> still as ghost—
>   with a tragic cry of recognition—
> "The Wheel" (N 209)

Already Graham identifies her visceral link to the play: she identifies with Clytemnestra. She will tell the story through (her recreation of) Clytemnestra's experience. Moreover, the dance will begin where the *Oresteia* ends, with Clytemnestra in Hades. Yet the dance will not move forward in time. Clytemnestra's question propels her back in time, to before the beginning of the first book of the *Oresteia*, namely "Agamemnon." In this meeting of ends, the dance begins with a "tragic cry of recognition": Clytemnestra sees, with Nietzschean resonance, "The Wheel," the wheel of eternal recurrence. Clytemnestra sees what Graham intends for her dance to communicate: the inner sensible substance or interior landscape of the textual drama itself.

Further evidence of the Nietzschean connection appears in a subsequent passage where Graham, citing a quotation from Eranos, introduces the image that later replaces the wheel for her as a sign of what Clytemnestra recognizes:

> From all the things that life offers us we have spun a net, a necessary cause that chains & enslaves us. But we retain the possibility of undoing this

Penelope's web, for we ourselves have woven it; once we have freed ourselves
from the servitude into which our actions lead us, we find ourselves at the
scene of our great task: possessing neither spirit or soul, we must achieve
a spiritual autonomy. (N 212)

What Graham sees in the story of Clytemnestra is what interests her in
religion: a moment of personal choice, an act of self-realization.
Clytemnestra is poised at the edge of her great task—the task of overcoming
herself and becoming who she is. As Graham studies the text of the *Oresteia*
searching for the point of entry—the "tragic cry of recognition"—she sees
the net, the web of flesh impelling Clytemnestra's action. In the second section of notes, Graham culls related images from the text: "women veiled in
dignities of black," "Net of death," "Fatal Nets," "Speech of Terror," and so
on. She sketches possible sequences of events, and then begins a line by line
study of the play, referring to commentaries and adding her own.

What Graham finds, using her physical consciousness to identify with
Clytemnestra, is a web of erotic attachments within whose conflictual
pulls Clytemnestra is caught. Graham perceives the physiological contradiction Clytemnestra herself endures—the contradiction that both necessi-
tates and catalyzes her rebirth. Graham reflects:

> Clytemnestra— . . . in the Underworld . . .
>> The endurance of return—of re-birth—
>> The private hell of a woman who has killed her love because her love
>> killed her creative instinct—
>>> —her child—(N 255)

Or then again:

> She is the glittering, regal, hate-breathing woman—
>> The woman whose creative instinct—her child Iphigenia—
>> has been killed by her husband—
>>> by her need for home,
>>> (her woman's nature has betrayed her)
> In a sense she is a "career woman" or a woman with creative gifts—in that part
> masculine in her strength of will & need to propagate her power—(N 259)

The conflict Graham perceives Clytemnestra living is physiological: it flares
among the multiple loves that are her body, the body of a woman.
Clytemnestra's love for her daughter pulls against her loyalty to her husband, which pulls again against her desire for her lover, and her consideration for her son. Her desire for children—for exercising her creative
instinct—has led her to bond with a husband who kills that creative
instinct. Moreover Clytemnestra feels this physiological contradiction as a
"private hell"—a situation in which she is betraying herself. What Graham
perceives in Clytemnestra's story is thus a tension akin to that experienced
by Graham as well as Duncan in feeling impelled to dance.

As Graham ponders the web of flesh in which Clytemnestra finds herself entangled, she again jots down images from the play itself: "nets of ruin," "tangled nets of fury." Then she describes:

How the interwoven lives & loves of four people weave the net which is the instrument of disaster—tragedy—
> It is the same net which catches Helen in its toils— laid upon her by Paris—
> The one used to ensnare
> Iphigenia—
> The one used to trap
> Agamemnon—
> The one used to bind
> Clytemnestra—
> The one used to entrap
> Orestes—
> Until some act breaks or disassembles or transforms it—it is a weapon of/
>   death—
> It is Karma—(N 259)

As Graham reads and writes about the text, she identifies (with) a web of words that ensnares, binds, and entraps the characters in the drama. It is the web of fate, or Karma—a web woven of the characters' own loves, their fleshed, gendered connections. Yet, in and through her visceral identification with Clytemnestra, Graham also perceives the possibility for an act that *breaks or disassembles or transforms* that net. The word web provides the occasion for such an act—and that act through which the web is transformed for Clytemnestra, Graham, and the audience is *dance*.

In the piece, Clytemnestra's act of (self) transformation occurs in her third round of remembering. As noted, in the first two rounds of remembering, Clytemnestra appears to herself as a ghost; she watches a visual image of herself dancing in the form of another dancer. One effect of these first two rounds is to stir a visceral identification between Clytemnestra and the audience: we are all, like her, watching her. In the third pass, however, Clytemnestra enters into the act of remembering with her body, recreating in and for herself the kinetic forms of her motivation. Already bonded to her, we enter her experience, seeing ourselves in her actions.

Graham's words in the *Notebooks* describe what occurs on stage in the moment when Clytemnestra decides to kill Agamemnon.

She sits facing Paul [Taylor as Aegisthus]
puts her hands on his face
He touches her r[ight] wrist
She bends him over to kiss her—she lies on chair—
    His head on her breast
    He puts her hand on knife

> She draws it—
> They stand—holding knife high
> She turns from him—
> He touches knife
> She walks away
> As she turns he sits with back to audience—face turned away from her
> (N 389–90)

What does not appear in these words (despite their vivid characterization of the encounter's sexual tension) is the intensity of the movements Graham created. Clytemnestra spends much of this scene in deep contraction, pulling into herself. In this visual and visceral sensation of inward pull, we discover with Clytemnestra what impelled her to kill Agamemnon: she was driven as much by her desire for Aegisthus as by her grief over Iphigenia's death.

It is this realization that allows Clytemnestra to give birth to herself and break the karmic net. She sees, in her own self, the physiological contradiction among her loves for husband, lover, and daughter and overcomes it. She does so by generating kinetic images of the tragic events of her life which allow her to affirm those events as expressions of her own lust for life. She affirms the power and pleasure of her own capacity for bodily creativity. As Graham notes, her interest was not to judge the morality of Clytemnestra's actions: "I did not make a moral issue of it—what mattered was not whether she was right or wrong but that she was vivid" (G73). Graham's concern was to discover the inner substance—the act of transformation that enables Clytemnestra to grasp and dissemble the net of flesh in which she is bound. In the dance, Clytemnestra affirms this net as offering her the occasion to exercise her bodily creativity; in making her dance, Graham affirms the web of Aeschylus' words as offering her the occasion to demonstrate how dancing allows us to experience the tragedy of the text as a stimulus to realizing the healing potency of dance.

In *Clytemnestra*, to conclude, we do not see a "danced version" of Aeschylus' play.[20] Graham's dance does not offer an interpretation of the drama; nor does it represent a retelling of the story from a "woman's" point of view. Rather, Graham's dance enacts a relationship of dance to the text in which the act of dancing serves as the practice through which the words of the text impact the audience. The movements of the dance communicate participation in Graham's experience of reading the text in and through the physical consciousness she has cultivated in practicing contraction and release. The kinetic images animated by Graham and her company members draw those trained as readers and writers to identify viscerally and visually with Clytemnestra, and in the process, to find their experiences of bodily being transformed. Graham creates an opportunity for spectators to experience Clytemnestra's fate as having the meaning Nietzsche and

Duncan describe tragedy as delivering: an affirmation of life. Audience members come to experience conflicting pulls of desire within themselves as the material out of which they make and become kinetic images of their selves, virtues, and values, their senses of nature and gender. As such the dance incarnates an alternative mode of valuing embodiment than that represented by Christian morality; it expresses a Dionysian faith in bodily becoming. Dance appears in this dance as the medium through which texts offer an opportunity for transforming the net of dichotomies that sustains the logical opposition between religion and dance, male and female, human and God, as lived in the modern west. In turn, by learning to dance, the dance suggests, we will be able to read even the most tragic texts in ways that encourage us to generate life-affirming values and move to incarnate them.

Errands and Acts

Clytemnestra is only one of nearly 200 dances that Graham choreographed over the course of her career. As such it is impossible to say that what I discern in this dance holds true for all of her dances. However, as one of her dances it does provide *a* perspective on what she was doing in making dances, and, in particular, in dancing texts. What it enables us to grasp is a relationship between dancing and texts that other dances share, if not as explicitly: dancing is theopraxis, a medium for generating and incarnating ideals of and about religion, when engaged in contested interplay with the production and reception of texts. Analyses of two other dances, *Errand into the Maze* (1947)[21] and *"Acts of Light"* (1981)[22] confirm this interpretation.

Errand into the Maze

There are two figures in *Errand*: "The Woman," and "The Creature of Fear." As the dance begins, The Woman is alone on the stage, back and to the right. Her crossed hands hover over her lower abdomen as a series of convulsions courses through her torso. She holds her body tightly against the pulsing waves, while her palms, turned to face her belly, roll with and against the swells. She strides stage left, back to the right, and with a fierce whiplash of torso and head, turns to face the audience. She enters a "maze," high stepping with crooked elbows along a rope laid on the ground. Following the rope she comes to what looks like a large "V," or the crux of two branches, and passes through. Isamu Noguchi describes this "doorway" he made for Graham as "suppliant hands, like pelvic bones, from which [and he quotes Martha] 'the child I never had comes forth, but the only child that comes forth is myself' " (Stodelle 1984:136). In the course of the dance, the man as The Creature of Fear enters three times. Each time The Woman must wrestle with him and push him away. After he enters for the third time, she pushes him to the ground where he lies, lifeless. She returns to the suppliant hands/pelvic bones, and moves between them. Standing in the crux of the bones, she lifts her right leg across herself from stage right to left. She steps through, giving birth to herself, walks forward, cups her hands in front of her heart, and sends them out into the space of the audience.

As with *Clytemnestra* so too with this earlier entry in her "Greek cycle." Graham reads this text, the legend of Theseus, with an eye to discerning the "act of self-realization" it describes. She looks to the text for what she intends her dancing to do: reveal a human. Here again, the fact that that human is "The Woman" is and is not relevant for reasons Graham may have learned from Duncan. In the dance, The Woman is revealed as a *human*, that is, as providing a lens on what human experience (and not women's as opposed to men's experience) involves. Graham thus honors Duncan's insistence that dance is able to serve women as a practice for exercising the process of their bodily becoming while avoiding any essentialist claims about the ideal of woman. In both dances, moreover, Graham adds to Duncan's account, for what dancing provides is an approach to reading mythic texts that reveals the dependence of the text's meaning on the gendering—the becoming woman or becoming man—of its characters.

In *Errand*, for example, the dance unfolds the myth of Theseus and Ariadne as a drama within The Woman's interior landscape. A program note from 1954 briefly narrates the myth then continues:

> Martha Graham's *Errand into the Maze* derives from this legend. But here the story has been transformed into a drama about the conquest of fear itself. The heroine enters a landscape like the maze of her own heart, goes along the frail thread of her courage to find the fear which lurks like a monster, a Minotaur, within her. She encounters it, conquers it, and emerges to freedom.[23]

The "transformation" described here is what occurs as Graham reads the text through the physical consciousness she has cultivated through her practice of contraction and release. Graham identifies viscerally with Theseus and recreates in her bodily movement the kinetic images of his self-realization. Her Woman will represent the human in Him; She enters the maze, kills the minotaur-like Creature of Fear, and finds a peace She shares with the audience. In making this dance, then, Graham poses a question similar to the one she posed to the *Oresteia: why* does the act of self-realization transpire? How is it that The Woman is able to break the hold of fear over her imagination? Her body? Her sexuality? Graham seeks to communicate participation in what the text cannot communicate: the *experience* of affirmation it describes.

Graham's inversion of the myth has led commentators to perceive this dance as proof that Graham uses bodies to express the world of the inner emotions: spectators witness fear and its conquest. However, in such a reading, the importance of the dancing lies in the *narrative* it provides: The Woman wins. The dancing does not appear to add anything to the myth. Such an interpretation rehearses the logic of the ascetic ideal, presuming a distinction between the inner (mental, emotional) and outer (bodily) self as the content of the dance's meaning.[24]

By contrast, when read in the context of the foregoing analyses of Graham's vision, practice, and dance making, what appears on stage in this dance is the process by which Graham, as a dancer, recreates a kinetic image of what appears to her as impelling the narrative. What appears in her dance, in other words, *is* a process of bodily transformation. A viewer does not experience an emotion, nor Graham's experience of an emotion, but rather Graham's recreation of the process in and through which a person transforms her experience of emotion. The dance reveals that what we might ordinarily consider to be an "inward" experience is something concrete, sensual, and informed by practice. From this perspective, the dance does the opposite of express an otherwise internal or psychological state; it operates to dislocate the ascetic ideal by demonstrating how our fear is a bodily phenomenon. It is because our inner life is not exclusively inner that it is open to transformation. In this way, the dance reveals the power of *dance* as a medium in and through which we can transform our experiences, such as those of fear. Dance appears as a way of participating in the ability of a body to overcome itself, to create beyond itself, to choose not to fall even when the rival wins. Dancing appears as a medium in which we come to know and affirm and negotiate the contested relationship between what we otherwise perceive as words and bodies, writing and dancing, or inner and outer expressions of self, as generative of who we are.

Graham's *Errand* enacts the power of dance to inform the inner self in several ways. For one, The Woman's convulsions do not act out fear; they provide an image of what it feels like to feel fear. When The Creature appears on stage, The Woman does not look at him. She registers his presence through the trembling contractions of her own body. Later, when The Woman conquers The Creature, there is no violent act. Holding on to The Creature's hands, while standing on his knees, The Woman draws strength from the fiercely concentrated arc of her contraction and slowly presses The Creature to his knees. She lets go and he falls. We feel the visceral pull of her movement, the effort it requires. Drawn to identify with her, we sense that the meaning of the myth is not its happy ending at all. The point of the conquest, when danced, is to communicate participation in the life-affirming sensation of creating over and against one's self—of becoming kinetic images of one's own making, including images of one's "inner self."

Further, the religious symbolism of the dance suggests that Graham was aware of the implications of her dance for a culture whose religious imagination is corralled by textual practices. Within Christian history the story of Theseus has been retold as the story of Jesus who descended in the maze of the underworld for three days, slayed the minotaur of death, and emerged to sit beside God in heaven. Medieval priests apparently danced along the labyrinths laid in the naves of European cathedrals during the Easter rituals. Across cultures, moreover, the labyrinth is a symbol of a spiritual journey. Whether Graham knew these details or not, the way in which her dance engages the myth suggests that she perceived her dancing as overcoming the contradiction lived by modern bodies between "pagan" and "Puritan" values; this dance is both. As both, the dance exposes whatever relationship we hold to our founding faiths as a function of how we participate in our bodily becoming. What the myth offers Graham's trained physical consciousness, in other words, is an occasion for enacting her self-creating, kinetic-image making potential—her potential to generate and become her highest ideals.

In her commentary on this dance, Graham adds a final twist that affirms this account of her work as revaluing Christian attitudes toward embodiment by communicating participation in a process of kinetic-image making. Graham explains how her physical performance of the dance serves to intensify the sense of inner space whose form the dance is often read as representing. *Errand* exists as a matrix of time and space in Graham's blood and muscle memories. Graham describes two occasions where she performed the dance in her mind in order to help herself through a close encounter with the fear of death.[25] In her mind, she envisioned the stage, the set, the costumes, and then imagined herself moving through the dance. Performing *Errand* in her mind allowed her to hold open a space within herself where she could gather the courage that was necessary for her to continue.

In this account Graham reveals how *Errand* in mind represents a space of resistance and responsiveness both to events in the world and eruptions of her own thoughts and feelings. This space is not imaginary because it arises in and through Graham's bodily practice and performance of a dance. She senses it. It is real to her. It exists as action of which she is capable; it is accountable to human bodies and to a community of spectators. As such, her use of the dance represents how the activities that humans practice influence their capacity to make and sustain alternative images to those which press upon them—images which, in turn, empty themselves in renewed thought and action. As Graham confirms, by rehearsing these kinetic images in her mind she was able to transform her experience, to become her self—to find as real for herself a sense of center (emotional and physical) in the face of acute stress. She finds for herself the peace and freedom that *Errand* represents The Woman as achieving. *Errand* thus serves to *create* an "interior landscape"—a space between inner and outer where such transformations of experience occurs. The dancing, even in mind, is effective not for its intellectual meaning or moral; nor for expressing an inner state of fear or its conquest. Graham knew the ending before dancing it in her mind. The dance proved therapeutic—seducing her back to life—by communicating participation in her own experience of bodily becoming. In recreating her dance in mind, she remembered her ability to create and incarnate kinetic images of her own transformation from fearful to affirming. She transformed her mourning into dancing by exercising the potential of dance as theopraxis.

"Acts of Light"

Nowhere is the sense of dance as theopraxis more explicit than in one of Graham's later dances, "*Acts of Light*." That this dance is considered atypical of Graham's work does not mean that it does not represent what many of her dances presuppose. In fact the structure of this dance is familiar: dancing reflects on itself. What is unique is that in this dance, the dancers *perform* the *practice* of her technique. As such, it enacts what Graham means when she quips: movement never lies.

In her introduction to a filmed version of this dance, Graham notes that "the most terrifying and beautiful prayer in the Bible" is "Let there be light." Although Graham attributes the title to a letter Emily Dickinson wrote to a friend, thanking her for her "gifts of light," Graham's choice of "acts" over "gifts" may represent this Biblical pull on her imagination. An act of light refers to a moment of creation, and in so far as that act is spoken, to the movement it presupposes. As such Graham's dance presents her practice as a recreated kinetic image of the movement constituting the act of creation described by the biblical word.

"*Acts of Light*" proceeds in three segments—A Conversation of Lovers, Lament, and Helios—which together cast a long trinitarian shadow. In the first segment, a man and woman glide through a stately, joyous duet between lovers, god and goddess. In the second, one woman wrapped in an elastic tube of white material, stretches and twists her shroud before being carried off by a group of dancers who hold her upright, arms outstretched, in the shape of a cross. The third section begins with a darkened stage, in silence. As the music swells, the audience becomes aware of bodies streaming along the outskirts of the stage, circling the center and closing in with percussive, straight-legged leaps. They are dressed in costumes designed by Halston—gold unitards which Graham describes as "golden raiment." In full light, the men and women assume their places on the stage, and begin the floorwork of a Graham technique class. As this section proceeds the company moves through a stylized version of an advanced class in unison, practicing the contraction and release in and through its major developments. From the floorwork they rise to exercises in the center, before pressing through space, across the stage, in clusters and lines, demonstrating the varied elements of Graham's movement vocabulary. As the dance winds to a close, the lovers distinguish themselves from the company, and reunite in a pose evoking sexual union. Surrounded by golden bodies, they consummate the instreaming of light. The company faces front, moves through kneeling positions, and settles into seated poses, with knees open, torsos pitched slightly forward, and straight arms, palms clasped, arcing down toward the earth.

In this dance, practice and performance come together in a vision for what dance is: participation in a body's ongoing act of (self) creation. As such, dance—specifically, the practice and performance of Graham's technique—appears as the act of light that illuminates the meaning of the Christian account of creation.

In the third section of this dance (which occupies the remainder of this discussion), Graham reenacts the moment of creation or resurrection. As noted, the stage is dark when this section begins; emptied of the crucified figure and her cohort. Then one man appears, center stage. His arms arc to soaring strings as the light slowly dawns. The line of dancers who stream onto the stage in golden raiment circle around him, as if responding to his call. They gather and settle to the ground like a flock of geese, poised for flight, ready to begin the practice of contraction and release—Graham's gift of light to the world.

The significance of this "act" appears on several layers. As is most obvious, the dancers are the conduits for the light that enters the stage space. They are themselves the rays of light—beautiful, golden bodies. The dancing of these bodies, then, is "of" the light in two senses: it is the movement that the light makes as it enters or is brought into the world. The dancing *is* the light. In a second sense, however, the movements of contraction and release that the dancers perform appear as the practices a person must or at least can perform in order to *become* light. Graham's dance does not celebrate a "natural" body; nor does she endorse conforming bodies to a codified ideal. The bodies that appear as ideal in this dance have become so by practicing the rhythms of breathing. They have become who they are by exercising a body's capacity to listen to, honor, and love itself; to create and become images of its highest ideals. As a result, the capacity to *be* the light appears as a function of a person's willingness to delve deep into the material of the self and do the hard physical work involved in disciplining oneself to the rhythms of one's own bodily becoming. "I use the words gods and goddesses principally, I think, to mean beautiful bodies—bodies that are absolute instruments" (G85, C1f).

Further, though the dancers are dressed alike and move in unison, they are of varying shapes and sizes. And there are many of them. The unison, in fact, accentuates how different each body is. The implication is that the acts involved in becoming human—or being the movement of creation—include learning how to move as one and as many, as individual and as human. Practice enables the realization of a community that shares in the making of light.

In these ways, the dance serves to transform audience members' experiences of the biblical prayer, "Let there be light." What was a verbal utterance now appears as the fruit of a practice in which people develop their physical consciousness of bodily becoming and exercise their capacity to make and become kinetic images of what they value. In so doing *Acts* implies that what the words of the Bible represent is not the existence of a bodiless, omnipotent Father, but rather the bodies of people creating beyond themselves. In the dance, these biblical words thus appear as occasions to *affirm* rather than deny our bodily participation in the movements by which light enters the world, and the world becomes real.

In sum, dancing, this piece reveals, is *movement made divinely significant*. The dance not only depicts the practice of dancing as the creative act that brings the world into being, the dance presents dancing as the action that defines what the divine or holy is. An implication, then, is that "I" who watch the dance am responsible for evaluating any and all (my) values based on whether and how they honor and enable this ongoing participation of human bodies in the creation of themselves and the world. Values that do so are *dancing stars* dancing through us. In short *Acts* suggests that "Let there be light" *means* "Let there be dance," for dance is the medium in and through which humans exercise their god-forming instinct in relation to themselves, becoming the gods and goddesses they are capable of becoming.

It is in this regard that Graham's statement "movement never lies" makes the most sense: movement participates in the coming into being of what that movement represents as true.[26] Graham's statement does not imply that there is a truth out there (as a signified) that movement (as a signifier) represents. Her point is that the act of performing dance is a moment in which an ideal of truth comes into being. Dance here, is what I call *theopraxis*. A person cannot *not* become what he dances. In this sense, movement *is* transparent to the conditions of its own making, but as a result, "the" truth forever remains a mystery, a becoming.

Nietzsche's Sun

In the Prologue to *Zarathustra* (chapter 2), Zarathustra compares himself to the sun—a golden body so strong, so radiant, that he has to overflow, spilling his rays across the earth. Zarathustra, like the sun, shines forth, sharing his light: his profound love for humanity as a process of its own overcoming. Zarathustra is a dancer, Nietzsche's dancing star.

Graham's *Acts*, in its reflexive performance of her dance practice, realizes the secret of Zarathustra's dance-like walk. The practice of contraction and release entails remaining faithful to the earth in the ways Zarathustra and Nietzsche describe: listening to the instincts, thinking through the senses, stirring and tapping energy. It is a practice that cultivates a physical consciousness of oneself as a rhythmic kinetic image maker, a bodily becoming. In doing so, such practice provides a critical perspective for evaluating the value of our values based on how they affect those rhythms. From the perspective provided by the practice of contraction and release, the body does not appear as a thing; it appears as the ongoing process of kinetic imaging through which we create ourselves, our worlds, and our gods. From the perspective provided by the practice of contraction and release, our freedom and agency appear as functions of our attentiveness to the conditions of our own health. From the perspective provided by the practice of contraction and release, it is possible to appreciate (Christian) antipathy toward bodily being, where it exists, as an expression of physiological weakness, and thus as a goad and guide to developing the kinds of strength we need in order to release our potential for the very gifts Nietzsche admits that Jesus promised: love and life abundant.

To the extent that Graham's work engages and advances Nietzsche's philosophical project, it simultaneously invites critical perspectives on that project itself. Revaluing values requires hard physical work. Inventing and enabling the kinds of physical–spiritual discipline that the success of his project requires will require a radical, ongoing revision of the forms of culture and the systems of education in which we (learn to) live. At every turn, allegiance to the ascetic ideal and its notions of truth, reason, and reality corrupt even the best intentions to honor and participate creatively in bodily becoming. There is even a risk that Nietzsche's philosophy and Graham's own dynamic practice will become idea-edifices, systematized, trademarked, and copyrighted.

In the end however, Graham's work distills from Nietzsche's a burning question that remains unanswered by him or her: what would it be like to live in a world whose forms and ideals, whose practices and values, were held accountable to realizing a vision for *dance* as an affirmation of life through movement?

Conclusion: Another Ideal

No artist is ahead of his time. He is his time: it is just that others are behind the time.

—Martha Graham in Armitage 1937:107

We have yet to catch up to Nietzsche, Duncan, and Graham. Although the world has changed dramatically since 1900, the year Nietzsche died, the critique of religion and culture he launched and the efforts by Duncan and Graham to engage and advance his vision of dance as a catalyst for revaluing values are as relevant as ever before.

As a culture, we are more capable of probing farther into space, reaching deeper into the recesses of the ocean, mapping thinner layers of conscious and unconscious thought, and manipulating bodily systems to a greater extent than ever before. At the same time, in every case, the increase in knowledge has made it easier to do what Nietzsche feared and Duncan and Graham sought to counter: to sustain faith in an other world, a world of abstractions. *We are unknown to ourselves, we men of knowledge—and with good reason. We have never sought ourselves—how could it happen that we should ever find ourselves?* (GM 15). We purport to believe in our senses and to honor our senses as the gateways through which we gather knowledge about the world. Yet we train our selves to ignore our senses by relying on our technological extensions of them to correct what it is we ourselves see, hear, taste, feel. We convince ourselves that technology can perceive, sense, and interpret to a greater and finer degree than our finite bodies. As a result, we remain immured in idea-edifices of our own making. Our knowledge, including and especially our knowledge of ourselves, is mediated to us through images—visual, aural, and kinetic—made for us by our own image-making technology. In search of knowledge, we have paradoxically surrendered the source of our own freedom—our bodily image-making power—to the technology we have made in our image. In doing so, we implicitly acknowledge what we simultaneously deny: that the image-making capacity of our bodily being is our greatest strength, and the characteristic that renders us human. *There are new things which have come into our consciousness . . . but we still don't know about our body* (G52).

Not only is it easier to avoid knowing ourselves and participating in our own bodily becoming, it is easier as well to escape from the instances of pain and suffering we do experience that might otherwise occasion knowledge of ourselves and others. We seek relief from physical, emotional, mental, and spiritual discomfort by sampling from a vast array of drugs marketed aggressively to us by pharmaceutical companies; by yielding to the barrage of entertainment options offered to us by media conglomerates; by relying on the cut and fix services sold to us by the medical community. It is easier than ever to consult "experts" who will tell us why we feel pain, and then to turn on the TV, pop a painkiller, dive into a carton of ice cream, or surf the web to help ourselves feel "better." When we are tired we drink caffeine; when we are hungry we snack on candy bars and soda. And every time we choose to sample the "drugs" offered by our culture, we support an increase in visual/aural noise so intense that it requires Herculean strength to create a space of silence and listen to the sensations of our own bodies. It is nearly impossible to discipline our consciousness to the rhythms of our desires— to eat when we are hungry and stop when we are full, or to sleep when we are tired and wake when we are not. We are trained by our culture, by our friends, colleagues, and our own selves to depend upon images of our body/selves mediated to us by our consumer culture as the source of our "salvation."

Meanwhile the forces of reading and writing on our religious and moral imaginations have never been more powerful. These forces ensure that the images we do consume are less accountable to our bodily becoming. Some philosophers of culture claim otherwise. They say that the word has become an image—a ray of light on a computer screen, immaterial, easily erasable, even transparent. We live in worlds of virtual images now. Yet this distinction between word and image masks how our image use reinforces the bodily practices that express and sustain the relation to our embodiment practiced and perfected in the acts of learning to read and write. To a greater degree than ever, people must learn to sit in order to think and work and write. People learn to arrest their bodily movement, dissociate their sense of self from their physical sensations, and maximize the mobility of their frontal limbs and lobes. Once we sat to think and write and then walked and skipped, ran and played, in order to recreate. Now we re-create by watching videos, playing video games, and instant messaging. In every case we reinforce patterns of bodily discipline that sustain the logic of the ascetic ideal. The values we generate, the thoughts we think, the choices we make express the physiological impoverishment of this relentless, constant education to ignore bodily becoming as a source of knowledge. Yet unless a person learns how to train her "body" into docility, she cannot function as a member of society. She does not fit, and the problem is hers. In short, the educational system has increasingly become a funnel for channeling all forms of intelligence and creativity into models of thought and action that sustain the hegemony of reading and writing as media for communicating and authorizing truth. *Present experience has, I am afraid, always found us "absent-minded": we cannot give our hearts to it—not even our ears!* (GM 15).

In the scholarly realms of philosophy and religious studies, these cultural developments have been reflected in an explosion of interest over the past 30 years in "the body" and its practices, rituals, and surfaces. Ostensibly, the concern among scholars is with retrieval: the body has been lost and must be found. However, the means by which scholars have approached that finding have served to reinscribe the conditions of the body's disappearance. Scholars treat the body as an object to deconstruct, a text to decipher, or a materiality to dissolve into social forces. Scholars thereby treat the symptoms of ignorance without addressing the causes.

Further, this new truth—that we are bodies, and that our bodies are inscribed—though touted for its ability to recover the body, has diverted the scholarly conversation into debates that reinforce the body's erasure as a generative source for knowledge. The codes written in, on, and as my body by social and cultural forces are scars, indelible tattoos etched in the flesh. I am marked for life, my identity established. I can write over my past scars but never come clean. Yet as Nietzsche grasped, such attempts to retrieve "the body" express the impoverished physiological conditions of scholars themselves. Our theories of self and body express the life conditions of persons raised to read and write, to have faith in the ascetic ideal, whether in the form of Christian values or scientific methods. Our theories reinforce our sense of bodies as living at a distance from where "we" are, and *we* must save *them*. As such, most approaches to finding the body fail to theorize a body's agency and creativity. We thus enforce our ignorance of the processes of bodily becoming in which we are participating by imagining the body in these ways. And in the process, we play right into the invisible hands of a consumer culture that needs to believe in selves who are impressionable, whose desires are always already colonized, who can be persuaded to buy and consume what they do not need. We become our ideals; our ideals become us.

Even the dance world has not escaped the reach of these forces. Dance has been domesticated by its twentieth-century successes. Now considered a fine art, though the least well-funded or attended among them, dancers occupy a rarified world maintained by their sheer love of the art and the deep pockets of donors. In attaining such successes, however, dance has lost its revolutionary edge. It is welcomed into the fold as a nonverbal art.

The effects on dance innovation and practice are noticeable. As Anna Kisselgoff reports in the *New York Times*, the dance world is currently in a place of quietude. Trajectories of creativity initiated by the early modern dancers have run their course. There are no longer great choreographer/ dancers reinventing the discipline as we watch. While Kisselgoff reminds us, with hope, that such periods of apparent stasis have preceded all explosions of dance innovation, she does not look more deeply at the causes of this lull, and what it expresses (Kisselgoff 2005:E1). In particular, this decline in dance innovation has been accompanied by a homogenization in techniques of dance training. What a person must be able to accomplish technically in order to be a "dancer" is increasingly defined by the codes and standards of ballet. As a result, nearly all dancers—whether modern, tap, or folk—take classes in ballet. While ballet has developed systems of training that honor the physics and physiology of different bodies, the goal is to develop strength, agility, and grace in executing a code of steps, much as Duncan described. Ballet *is* beautiful. It is a classical art. Its technique is effective in training the body. The lines and flow of extensions, the speed and height of the leaps and spins can be breathtaking. Nevertheless, ballet represents only one particular nexus of physiological, moral, and aesthetic values regarding what dance can be; and that nexus serves, in large part, to strengthen the idea of dance as a nonverbal, bodily art. Ballet aesthetics still abide by the logic of ascetic ideals.

Lost in the homogenization of dance training is what this book remembers: the sense developed by the modern dancers that the process of bodily becoming is a locus for discovering and exploring alternative forms of knowledge and alternative ideals of religion than those offered by the Christian values woven into the fabric of western culture. While the technique of dancers today is phenomenal, the elements of dance practice, choreography, and performance that Duncan and Graham called upon religion language to evoke are less evident. Few dancers practice their technique as a means for cultivating an inner responsiveness of mind and body. Few engage dance as a practice for cultivating physical consciousness, or tapping the power of kinetic imagination in order to participate consciously in their own bodily becoming and in the creation of cultural ideals. Today we teach and learn dance as an array of bodily signs that we can rearrange in various sequences to express whatever meaning we want the dance to convey—or not. For dance, we assume, need not be yoked to meaning.

In short, in the world of concert dance and its myriad schools and off-shoots, there is little sense of dance as offering its own logic of representation. There is little sense that dancing—learning to dance and practicing dancing—is necessary for becoming human, for becoming free. The world of dance training and performance has been assimilated into cultural dynamics that express and sustain faith in the ascetic ideal.

Taken together, these developments suggest the importance of Nietzsche, Duncan, and Graham for our time. By tracing the religious impulses threaded through their dance visions, practice, and performance, we find resources for discerning the dynamics of desensualization and oversensitization at work in our relationship to our own thinking, dancing bodies. Nowhere do we learn to love ourselves enough. The solution, as the foregoing analyses suggest, is not to cop a good attitude or form a healthy body image; or to lose five pounds or exercise daily. We must learn to be bodies differently. In the work of Nietzsche, Duncan, and Graham, we find guidance for generating and becoming ideals that affirm the value of bodily becoming, in all its radical implications.

The rest of this chapter offers an alternate ideal for what dance can be: a practice that helps us relearn our relationship to bodily becoming and reclaim the freedom rooted in our inherent creativity. It is an *ideal*, admittedly, and one that has not necessarily been realized in any person or place. Perhaps it will never be. As an ideal, however, its purpose is to not to impose itself on reality, but rather to guide those who are drawn to it in remaining faithful to the earth. It is an ideal that responds to the needs of the day as I/body live them. It is not an ideal that advocates reviving earlier forms of dance or religion; it represents a possible consciousness of dance that can inspire the creation of aesthetic forms and practices capable of countering the workings of the ascetic ideal. It is an ideal of dance in which the practice of dance gives rise to faith in the wisdom of bodily becoming.

The ideal is this: *Dance is a physical–spiritual practice through which we cultivate physical consciousness of our kinetic image making as integral to the process of becoming human. In so doing, we acquire knowledge we would not otherwise have, and a critical perspective on the moral and religious values, we incarnate consciously or not.*

Dance as Transformative Practice

I am sitting on a chair on stage. The audience settles into silence. The stage lights, glow. Music begins, and with it quotations by Descartes. I stretch, reach, twist, and curl as he lays out his method for imagining that he has no body, and for securing certain and true knowledge, especially of God. I begin a dance with the chair, winding in, around, and through the chair, pushing off and recoiling back, wanting to move off the chair, but not wanting to leave. I cling to it, allow myself to be cradled by it, and then surge out again, wanting my freedom but not wanting to touch the ground. I am living in a world in which abstractions are more real than bodily sensations, a world in which I am a thinker.[1]

Scholars of dance in the modern West tend to approach the problem of defining dance by beginning with the body. Dance is a human body in motion, rhythmic movement. Or dance is a particular use of the body for aesthetic rather than practical purposes. The process of definition then involves drawing lines, establishing categories, and in short, reinforcing the sense that dance is a nonverbal, bodily art, whose symbolic potential nonetheless aligns with models of representation predicated on verbal language. Yet as suggested in the previous discussion, such approaches to dance reinforce the logic of the ascetic ideal, and sustain the hegemony of the practices of reading and writing over the scholarly, artistic, and religious imaginations of those concerned.

In the light of examples offered by Nietzsche, Duncan, and Graham, dance appears differently—not as a body that moves, but as a transformative practice. As a *practice*, dance involves the repetition of a precise set of acts requiring the coordination of all faculties—mental, physical, intellectual, and emotional. It is a repetition, moreover, designed to *transform* and what is transformed is our relationship to what appears to us as our "body." Dance, in this sense, is a practice through which we become body and discover our bodies as the rhythm of their own becoming. It is a practice in which we encounter an experience and knowledge of ourselves as inherently creative.

Such wisdom is Dionysian in its paradoxes. In so far as we are becoming, we never properly are. In any moment, the movements we choose to repeat develop our potential for sensation, experience, and insight in one direction rather than another, opening and foreclosing possibilities available to us in the next moment. Our bodily movement is thus an *expression* of ourselves in the sense that it enacts this movement of our becoming. Our bodily movement also has *form* as determined by the range of potential strength and articulation our practice is drawing into physical consciousness. Thus, when idealized as transformative practice, dance appears as an activity in which we are always losing and finding ourselves at once, participating consciously in a logic of self creation and destruction. By providing us with a concrete experience of this rhythm, dance exercises our desire, willingness, and ability to become who we are.

The precise acts we practice as dance, then, matter to whether or not the practice of them effects this kind of transformation. The precise acts must themselves embody, in their kinetic forms, a deep respect and love for bodily becoming as a source of wisdom and intelligence, freedom and healing, creativity and love. To practice *dance*, according to this ideal, is to engage in the conscious repetition of acts that not only awaken a physical consciousness but guide its growing attunement to rhythms of bodily becoming. Only the practice of such actions is capable of facilitating the transformation in bodily experience we need in order to counter the effects of the ascetic ideals we have become.

It begins with breath (B 46). Such actions may begin with breathing. To dance by practicing the act of breathing, as Graham did, is to begin with the pulsating rhythm of each and every human life. Breathing is a kinetic pattern of tension and release, creation and destruction. It is the activity through which we recreate ourselves in every moment, throughout every day. To move in ways that honor and release the rhythm of breathing is to practice sensing what is happening in the present; it is to cultivate a stronger ability to make health-enabling decisions for our *life*. It is to enact a love for ourselves, a *holy selfishness*, which overflows to embrace all of existence.

In traditions of prayer and meditation around the world, breathing is a standard technique for clearing the mind, calming the body, and opening the heart. In such contexts, to focus on breathing is to provide oneself with a means to transcend the attachments and pressures of the day and float into another space of consciousness—one that is ostensibly freer, providing a person with a clear perspective from which to think and act when exiting the space of meditation.

In the vision for dance that Graham opens up through her engagement with Nietzsche and Duncan, however, a person breathes to *impel* rather than *still* bodily movement. A person listens to her breathing and recreates patterns of her breathing in order to discover new movement possibilities. She actively engages the muscular, mental pathways of energy that breathing takes and uses these patterns as the material, the rhythmic substance of her art, with which she discovers what her body can do. Moving while breathing, she uses the breath to open new spaces of sensation through which to move; breathing while moving, she uses the movement to open new ranges into which to breathe. The movements of the body and the rhythms of the breath push and pull each other into new shapes—kinetic images that are relative to an individual's body, level of facility, degrees of attention, honesty, and vulnerability. Dance *is* this push and pull between movement and breath.

With every breath we recreate ourselves. No two breaths are ever the same. One act of breathing transforms the body that breathes the next. Nor is breathing ever natural—especially when we aim to breathe naturally. With every inhalation we are *reborn to the instant*; we take in the food, oxygen, energy that becomes us. With every exhalation we *die small deaths*, emptying ourselves of the wastes and toxins, the lifeless matter we no longer need. Breathing is always relative (as is our thinking) to who we are (becoming) in the moment—to the specific practices and contexts in which we are exercising our senses. How we breathe is relative to what we are doing—to the actions we repeat willingly, under pressure, or by habit. Every breath we practice, in this sense, has a capacity we can choose to ignore: a capacity to draw our consciousness more fully into the ever-unfolding, sensory present.

At the same time, the way in which breathing is relative to our selves in any given moment is never given. While breathing is most often keyed to our levels of comfort, attention, and hope, our breathing need not express our state directly. What our breathing expresses rather is our relationship to the moment of bodily becoming we are. To practice dance as breathing in this way is to cultivate a *physical consciousness* of ourselves as always participating in a rhythm of creation and destruction, not by the fact of breathing, but through the act of engaging and recreating the significance that breathing has for us—the significance that our living has for us. When we practice breathing as the source of dance, we develop a physical consciousness not just of our becoming, but of how we participate bodily in the ongoing recreation of ourselves by virtue of how we practice breathing, consciously or not.

This participation and recreation, moreover, while physical, is never merely so. Our bodies are patterns of tension, past moments of breath crystallized into the habits and reflexes of bodily memory. Remembered breaths provide us with resources for responding to what another person says or does in familiar and expected ways. When we employ our stock responses, often unconsciously, we cement ourselves in patterns of physical and emotional tension we incorrectly identify as ourselves—envy, frustration, hatred, desolation, or despair. We can create out of such patterns. However, what we create from such emotional patterns most often functions to preserve them. We come to believe that our abstract worlds offer the only path to happiness. We learn to surrender and even despise the very creative agency our acts of value creation represent.

To practice breathing as a source of a transformative movement practice, by contrast, is to allow oneself to be drawn into the action of breathing—the movement of feeling and sensation. When we listen to our bodily movement through our breathing, we learn to sense where the energy is and where it is not flowing through our bodies. We feel where we are connected to ourselves and where we are not. We feel the levels of our energy, flush or not. We also feel pain as our desire to breathe collides with the bodily memories of past suffering. However, by posing obstacles to our moving/breathing, such patterns of emotional and physical tension teach us. The pain is not the problem; it is a guidepost to healing, showing us the dimensions of our selves whose becoming or self-overcoming we must encourage and allow.

We can do so by creating kinetic images of our breathing. When we discipline our physical consciousness to the act of breathing, we learn to shape and direct our breathing in new patterns, kinetic images. We learn to send breath to our muscles. We can invite past patterns of response to release and reform. We invite the wisdom of our bodies to recreate those patterns into kinetic images that enable the movement we desire. We thus experience our ability to transform pockets of fatigue and despair, tension and pain, into resources for dancing—into opportunities to revive the bodily wisdom we have learned to ignore. We regenerate our vulnerability to experience. *We heal ourselves*, not by returning to a place we once were. We are never the same. We heal ourselves by participating in the creative becoming of our bodily selves and transforming whatever happens to us into an occasion for knowledge, love, in a word, *bodily movement*.

Breath is the movement of our freedom. To practice breathing as dance is to learn how to cultivate the potential for freedom inherent within our capacity to make and become kinetic images of our breathing, beating selves. The freedom to respond in and to the moment is not a function of detached observation and rational calculation; it represents an ability to sense, engage, and meet the emerging challenge. It represents a willing participation in bodily becoming. Thus, if we take breathing for granted as given, we sever ourselves from the sources of our freedom alive in the present. We bind our spirits. Conversely, if we practice breathing as the source of dance we provide ourselves with a perspective—a physical consciousness—from which we can find in our breathing movement the wisdom of our own intuition. *Perhaps what we have always called intuition is merely a nervous system organized by training to perceive* (G41, 180).

To practice breathing as dance is, in all these ways, to revalue the relationship between mind and body, or spirit and flesh, which we have become in the process of learning to read and write, honor ascetic ideals, and believe in science, to whatever extent we do. When we practice breathing in movement, we practice releasing our bodily wisdom to move us in ways that express a generous, overflowing love for ourselves.

This idea is terrifying to those trained in Christian morals, among others. Often, people raised in Christian contexts view the body as a caldron of unruly desires that demands utmost vigilance and control. Those who espouse these values, consciously or not, believe they must use their will power to hold themselves in check. Otherwise they will wreck havoc on their relationships, their communities, and their worlds. They will sin against each other and God by putting their needs above those of others around them.

Yet this way of thinking about human beings is itself an expression of a physiological state, and of one that Nietzsche would describe as impoverished and even decadent. Our bodies do need guidance and discipline. However, the discipline I/body need is one that I/body create in and through attentive participation in my bodily becoming. When we attend to our breathing as we move, our bodily wisdom begins to move us with subtle, gentle impulses more refined than those provided by our rational thinking. Our bodily wisdom makes millions of tiny adjustments—a joint releases and stretches against itself; a limb falls into alignment, pockets of untapped energy flood into consciousness. We find ourselves moving with greater range and elasticity, less pain, more presence. We find ourselves responding to challenges with additional resilience and resources. We feel joy. The bodily wisdom released through the practice of breathing brings our thoughts and feelings and actions in line with the conditions for what Nietzsche called our *great health*.

In sum, dance is a transformative practice when the movements we repeat guide us in developing a physical consciousness of the rhythms of our breathing, and of how our breathing is a practice in and through which we release our creative potential, affirm our freedom, empower our healing, and recreate the conditions of our full flourishing. We need to discover such movements.

Our bodies are conscious, they are our consciousness; they are the consciousness in and through which we exist. To honor and embrace this fact by practicing breathing as dance is to open ourselves to new resources for imagining futures that address the suffering and injustices of the present.

Dance as Knowledge

What can I do? How can I move? I am on the floor, rolling, sitting up, reaching my toes into the air, doing what I have learned to do on the chair around the chair. Nothing. I collapse. Fetal position. Face to the wall.

Yet, I cannot rest. Something pulses within me, wanting to get up, to move, to live. I somersault to my feet, knees pressed together, wrists crossed in front of my body, I stand, weakly. I reach with a foot into space, lose my balance and catch it again. I fall into a circle and allow my scampering feet to draw back under me. As I press fruitlessly against knotted knees, I stamp the ground in anger. I try again, and dissolve in frustration. Another time, and my anger teaches me. In its pounding rhythms I feel a source of strength. I no longer need the chair.

I stretch my arms up into space. I find my hips and roll them left and right. I find space under my elbows, pressing them open, one by one. But then an elbow comes crashing down into my belly, propelling me back to the back of the stage. I internalize that blow. I repeat it, use it, and find a new arc of movement and then another. My freedom is etched by the costs of learning to move. The music picks up and although my knees are still locked, I find power within my constraints. I use those constraints, to make something beautiful. I am dancing my dance. I tip over the chair where I began, and fall.

If a person succeeds in creating beauty out of oppression, sickness, or tragedy, is the suffering thereby justified?

—On Fire

To dance as a transformative practice yields knowledge that is otherwise unavailable to us. First there is the knowledge of vitality, a vitality that comes with being in conflict with oneself. The practice of breathing in movement animates a sense of tension within physical consciousness. A body is always twisting against itself, lifting up from itself, reaching out from itself; pressing into itself, becoming different from itself. The rhythmic waves of our eating, sleeping, and lovemaking each follow a different amplitude and frequency; they cross, collide, amplify, and counter one another. Our instincts war against one another. Yet, because we are always working and moving in opposition to ourselves, we have a desire and a need to express ourselves constantly, and recreate ourselves anew. Under the influence of the ascetic ideal, we do so by creating models of the self or visions of an afterworld in which the conflict ceases and clear, cool reason prevails. Under the influence of dance, by contrast, we may learn that this conflict is the very source of our energy; it is what impels us to create beyond ourselves. It is the rich material out of which we create who we are. In providing us with a physical consciousness of our own vitality, dancing gives us a perspective of strength from which to engage and test our desires, discern their value, and follow the wisdom they represent.

In knowing this vitality, we know our responsivity as well. As we practice exercising the creative tension between breathing and bodily movement, we feel our own capacity to respond to cues, whether internal or external, with movement of our own making. Sensing our own difference provides us with resources for discerning and responding to differences in and among others. We know that we can find responses by *moving* differently—by imitating gestures, and recreating in our own physical consciousness the shapes of thoughts and emotions as they appear to us. We can imagine alternatives because we are already alternative to ourselves.

Further, the practice of disciplining our movements to the rhythms of our breathing generates a third kind of knowledge that Nietzsche, Duncan, and Graham all agree is essential to the project of revaluing Christian values. We know our capacity to recreate ourselves along the lines of our own healing wisdom. This wisdom is not in the body in a sense that it is coded in the flesh; we cannot access it through rational reflection. It is a wisdom that develops over time *as* the body. It is a wisdom that develops in relation to what we practice as we listen (or not) and experiment (or not) with the conditions necessary for our own flourishing. It is a body-specific wisdom that we encounter only to the degree that we risk having faith in our own bodies.

Specifically, the practice of dancing can teach us how to move in ways that express love for ourselves. This bodily wisdom is not theoretical (or capable of appearing as a formula); it is not practical (in the sense of determining moral action); nor is it technical (or able to guide us in accomplishing some task). It is kinetic; it exists in the movements it makes possible. While it can take infinite forms—word, image, gesture, sound—there is no form it must or cannot assume. At the same time, this bodily wisdom is never exhausted by its forms for it exists in the body as a sense of how to discern which forms comprise the best conditions for our ongoing health. This bodily wisdom is difficult to attain. We can acquire it only in and through practice. And even then we can never know "it" in the way we have been taught to know facts. It is a wisdom that is always becoming true and is never established once and for all.

To practice dance as breathing and breathing as dance, then, provides a person with knowledge of his bodily becoming as itself a guiding wisdom. He knows that his sensations and experiences communicate to him who he is (becoming). These experiences are the material of his self with which he can and must work in discerning and realizing the conditions for his ongoing creative satisfaction.

In sum, the knowledge dance offers of our own vitality, responsivity, and bodily wisdom allows us to engage the experiences of our lives differently. It provides us with the willingness and the desire to find in whatever happens to us the conditions for our own flourishing. It impels us to transform our experience of what has happened into moments we embrace as necessary. We find centers within ourselves. From the perspective opened by this knowledge, we are able to change our mourning into dancing.

Dancing Values

The knowledge that dance as a transformative practice can provide concerning our vitality, responsivity, and bodily wisdom is critical for revaluing values predicated on the ascetic ideal. This lived knowledge—and not reason per se—is what provides us with the perspective from which to challenge any abstraction, any ideal, or any god/dess, as morally wrong for the ways in which it does or does not promote our (great) health. In other words, such knowledge, while dynamic and ever-evolving, provides us with criteria for assessing the value of our values. That guiding criteria, fundamentally, is love: a book or person or ideal that dances is one that expresses and encourages a relationship to bodily becoming characterized by attentive, honest participation. This process of evaluation, then, is simply impossible without a physical consciousness of the conditions needed for our own health; it depends on having a physical consciousness of participating in the process by which we are always re/evaluating those conditions, learning through experience, and remaining open to the ever-unfolding present. For this reason, it is not an exaggeration to say that *every move we make expresses a moral universe.*

In sum, in practicing a dance that breathes and a breathing that dances, we acquire knowledge of the fact that we *must* practice if we are interested in revaluing the ascetic ideal where it appears in western civilization. Revaluing ideals demands hard physical and mental work because ascetic ideals have become us; ascetic ideals live even in our attempts to resist their effects on us. It is impossible to refute an ascetic ideal: any refutation rests on grounds the ascetic ideal sustains—faith in reason and truth. The only way to counter the ascetic ideal is to revalue it—that is, to learn how to use it and its effects as resources for enabling our ongoing bodily becoming. We must develop perspectives within ourselves from which to affirm the ascetic ideal as a human creation and as an occasion to extend our knowledge of human bodily becoming. We must learn to affirm the ascetic ideals working in us as moments in our own rhythms of self-creation—moments that can and need to be balanced by practices that encourage us to engage and exercise the freedom and creativity of bodily becoming. Only by developing this tension with ascetic ideals in ourselves, can we develop the internal rhythmic perspective from which to appreciate and live self-discipline as an act of self-love. Dance serves as a physical-spiritual practice that develops in us the capacity, courage, and willingness to cultivate such faith in bodily becoming.

The selves who we are are moving targets. To know ourselves as Nietzsche, Duncan, and Graham advocate is a life-long process of immersing ourselves in the moment and allowing ourselves to be challenged, grown, expanded, and exercised in response to whatever experiences arise. We can never know ourselves exhaustively or definitively. We never *are*; we are to ourselves an ever-receding horizon of possibility. For Nietzsche, this fact is the source of our joy, our freedom, and our love for life. For Duncan and Graham, this fact is what makes dancing—actually learning to dance—both necessary and desirable for human becoming.

* * *

Today "dancing" and "religion," as conceived and lived in American culture, represent two faces of the ascetic ideal. Members of the dance world perfect their physical technique; practitioners of religion seek spiritual perfection in this life and the next. The two meet warily from time to time, on stage or in a religious setting. Yet, in so far as we perceive and honor an opposition between them, we are not free. We lack the deep spiritual freedom to transcend—not "the body" but the shapes of the past ideals we have become, in body and mind. We still believe that if we can think the right thought, or put our minds and hearts in the right place in relation to a divine, we can find the resources in ourselves that we need in order to love. Yet such beliefs foster styles of dependence; we grow incapable of mustering the strength to respond to the challenges of life without engaging in acts of self-denial. We respond to our pain by hurting ourselves. We beg for forgiveness and seek a sense of our oneness with the powers that be, even in their apparent opposition to us. We thus dwell inside idea-edifices that express our faith in the ascetic ideal.

We are ideal makers. We cannot not be. There is no material reality on which to lean; no spiritual reality into which to escape. There is only the process of our own becoming-body in and through the ideals we generate. It is our responsibility to take up the challenge of generating and becoming ideals and bodies that guide us in overcoming the life-negating effects of the ascetic ideal.

In and through the transformative practice of dance, we need to reclaim the freedom lodged in our bodily becoming—to make choices that give us time to discover ourselves moving; that encourage us to listen to our bodily becoming attentively, that guide us in disciplining our consciousness to the complex and conflicting rhythms of our deepest wisdom. We need practice in being a becoming. Our reason is finite; our bodies are infinite. We are not free to become whatever our minds can imagine; we are free in any moment to move with who we are, in response to what we perceive, and to become who we are and will be. We need to ask ourselves: *How must I dance today in order to dance again tomorrow?* It begins and ends with breath.

*   *   *

The body that writes is still, stiff. The mind that thinks is empty. Consciousness is concentrated into a point so small that it disappears. The writer has thought herself into a void. It is time to draw consciousness back into the fibers of the body, and breathe blood back into the reservoirs of muscle memory. It is time to dance, so that tomorrow I may write again.

"And one . . ."

Notes

Introduction: Reading Nietzsche's Images of Dance

1. John Atwell has written one article, "The Significance of Dance in Nietzsche's Thought," in *Illuminating Dance: Philosophical Explorations* (Ed. Maxine Sheets-Johnstone, 1984). Often "dance" appears in the title of an article or book even though the piece contains no sustained analysis of Nietzsche's dance imagery. Tyler Roberts, in *Contesting Spirit* (1998), asks how dance as a figure helps Nietzsche conceive his philosophical project, but does not follow this question with an analysis of Nietzsche's dance images (34, 46). Roberts' concern is to demonstrate how close Nietzsche is to the Christianity he appears to reject: he conceives philosophy as a "the spiritual practice *par excellence*" (17, 76). In the process Roberts reduces the dance of free spirits into a "dance of the pen" (156), and then again interprets that dance of the pen not as a kind of "self-cultivation closely related to traditional forms of religious ascetic practice" (78), but rather as a way of writing with "a certain insatiable desire, abandonment, and creativity on the limits of his philosophical discourse" (128). In associating dance with this ecstatic state of mind, Roberts loses the critical perspective on writing as a practice of embodiment that dance as a metaphor provides. As a result, he misses an opportunity to understand how Nietzsche's proximity to Christian values is what enables Nietzsche's radical critique and revaluation of them. See chapter 1 through 3.

2. Kaufmann's translation of *The Gay Science* (1974) is an exception. In a footnote to aphorism 327 he acknowledges Nietzsche's dance imagery as participating in a cluster of images including "light feet" and the "spirit of gravity"—images Kaufman interprets as critical for understanding how "gay science" does not involve repudiating science, but practicing it differently (GS 327n). In the *Portable Nietzsche* he writes, "The dance is to Nietzsche a symbol of joy and levity, and the antithesis of gravity. He associates it with Dionysus" (1954:118).

3. For a concise overview of Nietzsche scholarship, see Bergoffen (1990). See also, "Introduction to The Cambridge Companion to Nietzsche," (Magnus and Higgins, 1996) and the selected bibliography in the same volume (385–97).

4. For an overview of feminist interpretations and interpreters of Nietzsche, see Kelly Oliver and Marilyn Pearsall, eds. (1998), and Peter Burgard (1993).

See also Kelly Oliver (1995), Luce Irigaray (1991), Jean Graybeal (1990), David Krell (1986), Jacques Derrida (1979). Those interested in Nietzsche as a religious thinker include Jean Graybeal (1990), Tyler Roberts (1998), Tim Murphy (2001), and the contributors to Weaver Santaniello's collection, *Nietzsche and the Gods* (2001).

5. For the history of dance in philosophy and religious studies see Sparshott (1988), LaMothe (2004).

6. Derrida's most extended reflection on dance metaphors appears in his discussion of Mallarmé in "The Double Session," in *Dissemination* (1981:173–285). He also uses dance to figure the relationship between the sexes in "Choreographies" (with Christie V. McDonald, in *Diacritics* 12(2): 66–76).

7. The effect of this effort, metaphorically speaking, is to superimpose the character of writing as a discursive practice onto these nondiscursive forms, such that all forms of communication appear as texts that can be read—texts whose meaning is, in the end, undecidable. See Derrida's *Of Grammatology* (1976).

8. For further discussion and notes on biographical information, see chapter 4.

9. For additional discussions of Duncan and Nietzsche, see LaMothe (2003, 2005a).

10. For further discussion and notes on biographical information, see chapter 6.

11. Stodelle (1984) reports that this sentence was quoted in W. Adophe Roberts, "The Fervid Art of Martha Graham," *Dance Magazine*, New York, August 1928. For Stodelle, what interested Graham about Nietzsche was his "romantic fervor" and his understanding of "creative compulsion" (37–8).

12. For a recent exception, see Janet Lynn Roseman, *Dance was her Religion: The Spiritual Choreography of Isadora Duncan, Ruth St. Denis and Martha Graham* (2004).

13. For discussions of these developments, see Banes (1987, 1993) and Foster (1986:chapter 4).

14. For discussion and debate of these developments, see articles in *Researching Dance: Evolving Modes of Inquiry* (Eds. Sondra Horton Fraleigh and Penelope Hanstein, 1999); *Routledge Dance Studies Reader* (Ed. Alexandra Carter, 1998); *Meaning and Motion* (Ed. Jane Desmond, 1997), and *Choreographing History* (Ed. Susan Foster, 1997). Susan Foster is one who develops an approach to dance as "bodily writing" (1997:3–21). Jane Desmond, in "Embodying Difference: Issues in Dance and Cultural Studies," argues that we need to enlarge our understanding of "bodily 'texts' " to feature dance in all its forms as a "primary not a secondary social text" (Carter 1998:160). We need expanded understanding, Desmond argues, of "the ways in which the body serves both as a ground for the inscription of meaning, a tool for its enactment, and a medium for its continual creation and recreation" (161). At the same time however, there are those who voice concerns with the limitations of the text-based paradigm. As Marcia Siegel notes, for example, in "Bridging the Critical Distance": "We in this country have almost no knowledge of dance as ritual, dance as a spiritual lesson, dance as a historical memory, dance as a means of communal celebration—or at least our arts pages don't recognize them" (Carter 1998:96).

While Siegel is addressing the challenges of working across cultures, I would argue that the same limitation hinders our understanding of American dance traditions as well. Fraleigh is one scholar who is developing an alternative that parallels and inspires my work here. In "A Vulnerable Glance: Seeing Dance through Phenomenology" (Carter 1998:135–43), *Dance and the Lived Body* (1987), and *Dancing Identity: Metaphysics in Motion* (2004), Fraleigh draws upon the field of existential phenomenology for resources in studying dance as a lived experience.

15. See Ann Wagner, *Adversaries of Dance* (1997).

16. I am aware that to some, this project will not appear as "philosophy" at all. It calls into question the privilege philosophers award to written expressions of truth and the authority they accord to rational thinking as a medium of gathering, storing, and communicating knowledge. However, I do it with the aim of strengthening our practice of writing and our use of reason, by holding them accountable to the physiological conditions they express and perpetuate. As Grace Jantzen affirms: "That a feminist philosophy of religion which seeks a new symbolic and social order requires an altered demarcation of the boundaries of the philosophy of religion is not an objection, it is part of the point" (1999:264).

17. Specifically, in her reconsideration Jantzen details how representatives of this Anglo-American tradition fail to appreciate embodiment as enabling condition of knowledge (30). They presume a *desire* for rational thinking, for certain and true knowledge about God, that they intellectually dismiss as irrelevant and even harmful to their pursuit. As Jantzen writes, "Without the intense desire to be rational, and indeed to construe rationality along the lines of justified beliefs about which one can formulate belief policies which in turn preserve and defend such rationality, the whole enterprise could not get started" (84). In response, Jantzen offers expanded notions of "reason" and the "symbolic" that integrate consciousness of the desire, imagination, and embodiment she finds implicitly at work in the Anglo-American philosophy of religion. She argues for "a wider understanding of reason that includes sensitivity and attentiveness, well-trained intuition and discernment, creative imagination, and lateral as well as linear thinking" (69). "Embodiment" Jantzen affirms, "is the very site of transcendence" (252). Finally, Jantzen's reconstruction implies a *performative* dimension (225). This dimension represents the necessary condition for ongoing theoretical activity, and for human flourishing in general. To flourish, humans must engage consciously and thoughtfully in the ongoing *becoming* of their own bodily being in thought and action (254).

18. I will argue that the dances of the modern dancers are *not* nondiscursive: they do not exist in opposition to what counts as "discursive." Both dancing and writing engage, cultivate, and express a particular patterning of bodily becoming in the interests of communicating. However, in so far as Jantzen's intent is to affirm other media of communication than reading and writing as integral to the philosophy of religion, I move with her.

19. For further elaboration of this method, see my *Between Dancing and Writing: The Practice of Religious Studies* (2004), chapter 6.

Part 1 Friedrich Nietzsche

1 First Steps

1. In this passage "idealizing art" is the act of "showing worshipful gratitude" (HH #214, 128).
2. For biographical information see Ronald Hayman's *Nietzsche: A Critical Life* (1980); Walter Kaufmann's *Nietzsche: Philosopher, Psychologist, Antichrist* (1950); and Heinz Friedrich Peters' *Zarathustra's Sister: The Case of Elisabeth and Friedrich Nietzsche* (1985). See Hayman (1980:chapter 1–3).
3. Based on a letter from Elisabeth to Friedrich, cited in *Zarathustra's Sister: The Case of Elisabeth Forester Nietzsche* by Heinz Friedrick Peters (1985:17).
4. See also: "To what then shall I compare the men of this generation and what are they like? They are like children sitting in the market place and calling to one another, 'We piped to you, and you did not dance; we wailed and you did not weep'" (Lk 7:31–32).
5. See Psalms 149 and 150 in particular: "Praise him with timbral and dance" (Ps 150:4). For the many other references to dance in the Hebrew Bible and New Testament, see *Dance as Religious Studies*, (1990). See also LaMothe (2004:Introduction).
6. Kaufmann reports that even Emerson, a favorite of Nietzsche, links dancing to poets and professors. Kaufmann cites the following passage from Emerson, admitting that he has no proof that Nietzsche actually read this lecture of Emerson's: "I think the peculiar office of scholars in a careful and gloomy generation is to be (as the poets were called in the Middle Ages) Professors of the Joyous Science, detectors and delineators of occult symmetries'—and so forth, down through 'music and dancing' " (GS 10).
7. See Jowitt (1988:chapter 1).
8. Thomas Brobjer, "Nietzsche's Changing Relation with Christianity," (in Santaniello (2001:137–57)) is one who describes Nietzsche's "loss of faith" as precipitated by his reading of Feuerbach in 1861. By 1865, Nietzsche had stopped studying theology.
9. For the classic expression of Feuerbach's thesis, see *The Essence of Christianity* (1957).
10. Nietzsche's complex and multi-dimensional critique of Christianity unfolds in greater detail throughout this chapter and the following two. Basically, Nietzsche holds that Christianity posits ideals of human being that encourage humans to think of themselves as souls or minds in bodies in such a way that they neglect their earthly existence—their health, sensory experience, bodily becoming—in pursuit of rewards to come in an afterlife or in relation to a supernatural being. While Nietzsche's critique may not do justice to the range of historical Christianities, it is powerful in relation to those branches of Christian religion—mostly modern Protestant—that do demonstrate the characteristics he describes.
11. Brobjer (in Santaniello 2001) claims that Nieztsche was not occupied in his early years with Christianity "as a major philosophical or cultural problem or opponent" (147–8). Regarding *Birth* he writes: "Christianity and related

themes are completely absent" (143). I argue that Nietzsche's relationship to Christianity was always his concern. His challenge was to establish a critical perspective on it—one that would enable him to acknowledge his deep indebtedness to Christianity for his very ability to resist it. See chapter 3.

12. The logic here closely resembles Kant's description of a person in the face of the sublime. For Kant however, the experience of the (dynamically) sublime produces a mental response, in which a person affirms himself as having the rational capacity to transcend such threats to his bodily being: "Sublime is what even to be able to think proves that the mind has a power surpassing any standard of sense" (1987:#25, 106). For Nietzsche, the experience, while having a mental component, is physiological; it involves a rise in energy, a change in capacity for perception, the opening of a new perspective within one's self based on knowledge of one's bodily being as a process of becoming. See Kant's *Critique of Judgement* (1987), Book II: Analytic of the Sublime, especially #23–28.

13. Music can do so, Nietzsche insists, by embodying a tension he discerned earlier in the relationship between music and dance. He writes, only through the "wonderful significance of *musical dissonance*" can a spectator know directly the "primordial joy experienced even in pain" that both the music and the myth represent (K1, 152; BT 141). Dissonance allows spectators access to this knowledge by being a simultaneous sounding of different notes that amplifies the tension between them as the source of beauty. The tension within the music allows spectators to have an experience of a conflicted unity—a unity in which the elements are not homogenized but set against one another in ways that heighten the energy and intensity of the moment. As such, musical dissonance enacts "the playful construction and destruction of the individual world as the overflow of a primordial delight" (K1, 153; BT 142).

14. Nietzsche elaborates the importance of this shift for him: "What reached a decision in me at that time was not a break with Wagner: I noted a total aberration of my instincts of which particular blunders, whether Wagner or the professorship at Basel, were mere symptoms. I was overcome by *impatience* with myself . . . it became clear to me in a terrifying way how much time I had already wasted—how useless and arbitrary my whole existence as a philologist appeared in relation to my task" (K6, 324; EH 286). He calls *Human* "a monument of a crisis" (K6, 322; EH 283).

15. Although Murphy (2001) discusses the "nerve impulse" source of metaphor and the "generative loop" that resists any foundational (material or ideal) ground, he does not tease out how concepts inform sense experience—only how they inform our interpretations of sense experience (chapter 5). Thus, he misses the heart of Nietzsche's critique of metaphysical dualism—namely, that it encourages us to perceive ourselves as minds in bodies, to ignore the education of our senses, and to live through concepts. It is not just, as Murphy says, that the religious body and religious beliefs construct one another; a "body" is precisely that which has it own livingness, which resists construction, because "it" is the movement of its own creative, context-specific becoming. Nietzsche, I argue, wants to reorient us to this livingness.

16. See: "The more volcanic the earth, the greater the happiness will be—but it would be ludicrous to say that happiness justifies suffering per se" (K2, 339; HH #591, 249).

2   Free Spirits

1. See Hayman (1980:chapters 6 and 7).
2. As he confirms: "My *gaya scienza* belongs in the interval and contains a hundred signs of the proximity of something incomparable; in the end it even offers the beginning of *Zarathustra*, and in the penultimate section of the fourth book the basic idea of *Zarathustra*" (K6, 336; EH 296).
3. Commentators tend to treat these books separately, and in relation to *Zarathustra*, make arguments about why it should "count" as "philosophy." Referring to such discussions, Gooding-Williams notes in his book *Zarathustra's Dionysian Modernism* (Stanford University Press, 2001): "What is missing in this conception is the recognition that literature can contest an established philosophical diction" (12). Literature, he argues, can *produce* philosophical problems, even though, "As a productive act of the imagination it creates vocabulary, concepts, questions, and concerns that stand at a distance from the mainstream of Western philosophy" (13). By approaching the question through the image of Nietzsche's "daring dance" an alternative explanation arises, one that perceives the necessity of using two complementary styles in relation to the project of transforming our experience of God's death.
4. Gooding-Williams (2001) argues that the driving question of *Zarathustra* is whether or not it is possible to create new values. Zarathustra, the character, is "devoted to the project of creating new values," and as such he is a "modernist who, articulating his vision of the overman, aspires to create new, non-Christian-Platonic values that will transform European humanity"(5). Via *Zarathustra*, Gooding-Williams writes, Nietzsche raises the question of whether and how the creation of new values can occur. Gooding-Williams concludes that it can.

    Graybeal, in *Language and the "Feminine" in Nietzsche and Heidegger* (1990), relies on Kristeva's theories of language to interpret the driving problematic of *Zarathustra* as a problem of language: Zarathustra must learn how to speak (40). It is an: "extended meditation on the relation between silence and saying, between vision and verbalization, between despair or disgust over the inadequacy of language and the exaltation of liberated poetic flight" (40). From this perspective, dance figures in the poetic register as a figure for a way of speaking. See note 10.
5. David Krell ponders the question of why Zarathustra, as the tragic hero, never dies in his book *Postponements: Death, Woman, and Sensuality* (1986). However, if the death we must learn to experience as a condition for ecstatic bliss is God's and not Zarathustra's, the problem shifts. Krell himself describes plans in Nietzsche's journals for a Fifth Part of Zarathustra, with the title: "The Eternal Return/ Zarathustran Dances and Processions/ First Part: God's Wake" (78). Zarathustra, as a singer and dancer, is more closely linked to the chorus of Attic tragedy than to the tragic hero of the dramatic narrative. He lives on in the elemental rhythms he enacts.
6. For a prominent example, see Gooding-Williams (2001). Gooding-Williams argues that "[T]he tragedy of the tightrope walker signifies the death of the

possibility of overcoming man" (94); and that in this moment, "Zarathustra recognizes that his descent, regardless of his intention, may do naught but leave humanity suffering the experience of meaninglessness" (94). As such it embodies the central problem of the text. See note 4.

7. Kathleen Higgins is one who comments on the significance of this fact. See *Nietzsche's Zarathustra* (1987:86–7).

8. While Gooding-Williams discusses the need for bodily resurrection or transformation in depth, he is less clear about what that transformation involves or how it occurs. As he writes: "The will to power that causes itself to revalue the passions is the will to power of the child: it is the self-reflexive will to power of a spirit that . . .'wills its own will' " (127). The child, saying "yes" to the game of creation "is the transformation of passional chaos into a newly integrated body" (127). I argue that the images of dance add another level of interpretation to the "self-reflexive" will of the child and thereby provide needed understanding of the transformation as internal to a body's own rhythmic becoming.

9. Parkes (in Santaniello 2001) is one among many who reject the stereotypical view of Nietzsche as advocating an overflow of unbridled Dionysian passions (115). He acknowledges in Nietzsche "an appreciation of the vital power of the emotions and a refusal to let that power be lost through the reduction or extirpation of affect" (130). However, by focusing on the "emotions" Parkes misses the opportunity to acknowledge the role of physical discipline in educating humans to the self-creating logic of their bodies.

10. For an example of the kind of interpretation I intend to complement, see Graybeal (1990). Graybeal interprets the dance metaphors in "On Reading and Writing" as images of poetic language. She writes that Zarathustra "achieves through *semiotic* pleasure the flight of ecstasy he seeks, a separation from himself, from the self he leaves below and behind, the one that is serious, solemn, profound, and earthbound, constrained by categories and dead language" (45). For her, a god who dances is a "supreme image of a god who understands how to dance" (43); it is an image of poetizing, achieving semiotic release (43). While Graybeal attests to the bodily character of poetic language (for example, when writing that "the reader is led insensibly to an experience of this pounding, rushing course of language in the body, as if it were life's blood itself" (57), or in describing Nietzsche's effort in Zarathustra as creating "words and honors for the body and the earth"), she does not admit that the process of attaining this ability to demonstrate the jouissance of language might have anything to do with the actual practice of dancing. Moreover, it is poetizing aspiration, for Graybeal, that qualifies Nietzsche as a "religious thinker" (4). Thus her work implicitly reinforces an antithesis between the act of dancing and religious values, one that Nietzsche contests.

11. For a parallel reading of this passage, see Graybeal (1990).

12. Why in the form of a woman? This appearance raises the question of Nietzsche/Zarathustra's relationship to women. Derrida insists that the only "woman" that appears in Nietzsche's texts is an *image* of woman—and an image of woman that is un-determinable. While I agree that an "image" is all that appears, the dance imagery allows for a different interpretation of what that image implies. It implies that our ideal of woman is an expression of our own

bodily health, that we must claim responsibility for participating in the creation of ideals of woman that value the bodily becoming of those who identify them-selves as "women." This song introduces the problem of negotiating gender values in ourselves as central to the problem of learning to love humanity.

13. Compare this interpretation with Gooding-Williams (2001): "Zarathustra's metaphor for his reclaimed openness to life is Cupid's dance, a dance with girls who, when they first recognized Zarathustra, stopped dancing, as if they had sensed his allegiance to a will to truth that renders him a stranger to the danc-ing body's power to be moved by the uncreated passions that claim it" (168). Gooding-Williams argues that Zarathustra's "will to truth" prevents him from "going under" to the uncreated passions as he must in order to create new values. I argue, as noted, that the obstacle Zarathustra encounters is a contra-diction internal to his own love for humanity.

14. For a discussion and analysis of the two versions of *Ecce Homo*, see Jean Graybeal, "*Ecce Homo*: Abjection and the Feminine," (in Oliver and Pearsall 1998:152–69).

15. Roberts (1998), in an interpretation of "The Other Dancing Song," remarks that Zarathustra dances with life. Actually, he dances after her. Roberts inter-prets the dance as "a kind of answer to the problem of life—not an answer in the sense of a solution, but in the sense of a response" (130). That response involves letting go of the boundaries of the self such that "one is not oneself" (131, 133). Roberts contends that Zarathustra's realization, as an ecstatic expe-rience, can be figured "in a dance or a song of love—but it cannot be the stuff of philosophical doctrine" (133). Roberts later argues that Nietzsche subsumes the performance of "the real" in the act of writing. He writes that Nietzsche practices a "mysticism of metaphor" (162): he "seeks to communicate the joy of becoming by transfiguring the dancing of the Dionysian ecstatic into the danc-ing of the pen" (156). In these sentences, Roberts claims that the "dancing" refers "to a disciplined, deliberate practice of writing" (162).

As the foregoing discussion suggests, I read "dancing" as not only an image of self-loss, but as the rhythm of loss and affirmation through which a person creates his or her bodily self. As such, dancing does not figure (at) the limits of Nietzsche's philosophy but at its very core. Dancing as a figure enables Nietzsche to represent a kind of *bodily* participation in the flux of life that provides a critical perspective on writing as itself an aesthetic, bodily act.

16. A comment by Krell about Nietzsche is relevant here: "He knows that without the exchange and ringdance of male/female in him he cannot create" (Krell 1986:75).

17. As Debra Bergoffen writes in "Nietzsche was no feminist . . .": "in willing the eternal recurrence Zarathustra will need the complicity of the woman who speaks for herself. Without it the eternal recurrence runs the danger of becom-ing what Heidegger said it was: the last gasp of metaphysics" (In Oliver and Pearsall, 1998:232).

18. Nietzsche introduced this idea of going backwards in *Human*. There he insists that it is not enough to *understand* that metaphysics is an error, a free spirit must *go backwards* in order to understand the "historical and psychological justification

in metaphysical ideas" (HH #20, 27). If she does not, she will rob humanity of its greatest accomplishments. *Zarathustra* adds the idea of *willing* backwards in a way that makes the task creative and not just scientific. The aim is to transform our experience of those justifications, learning from them how to move forward.

19. Murphy (2001) interprets "revaluation" in terms of the one metaphor for metaphor he chooses, *ubertragen*, "to transport" or "to carry over." In doing so, however, he ignores what insights Nietzsche's dance imagery has to offer terms such as reverse (*umgekehrt*); move (*verlegt*); turn around (*umgedreht*) in relation to Christianity in particular. He interprets the creativity inherent in metaphor-making as one of creating relationships between two entities in different domains. His understanding of religion conforms to this model: religion, he writes, is a "dynamic and fluctuating ensemble of relations in struggles of ongoing cultural contact" (147). Yet such readings of metaphor and revaluation sustain the centrality of beliefs in religion. The movement between and among domains remains a mental act. By approaching revaluation of Christian values through dance imagery, it is possible to mount an understanding of revaluing religion that involves reeducating the senses to the rhythms of bodily becoming, and that acknowledges Nietzsche's insistence that it is impossible to refute or oppose Christian values because they live in and through our bodies. See chapters 3 and 9.

3 Loving Life

1. Genealogy: "the cause of the origin of a thing and its eventual utility, its actual employment and place in a system of purposes, lie worlds apart; whatever exists, having somehow come into being, is again and again reinterpreted to new ends, taken over, transformed, and redirected by some power superior to it; all events in the organic world are a subduing, a *becoming master*, and all subduing and becoming master involves a fresh interpretation, an adaptation through which any previous 'meaning' and 'purpose' are necessarily obscured and even obliterated" (K5, 313; GM II:12, 77).

2. *Will to Power* contains notes that Nietzsche wrote but did not publish between 1884–1888. Even though what it contains does not derive exclusively from 1888, it does reflect Nietzsche's post-*Zarathustra* sensibility. I do not privilege either the published or unpublished works as providing a picture of the "real" Nietzsche, but treat all of Nietzsche's writings as representing different perspectives on his thinking.

3. This idea of being a contradiction runs throughout Nietzsche's works, and comes to prominence after *Zarathustra*. In *Science* he notes: "[One] must first of all 'overcome' this time in himself—this is the test of his strength—and consequently not only his time but his prior aversion and contradiction *against* this time, his suffering from this time, his un-timeliness, his *romanticism*" (K3, 633; GS #380, 343). However the idea of a *physiological* contradiction seems unique to the works of 1888.

4. I am not arguing here that Nietzsche is closer to Christianity than he himself allows (See Roberts 1998). He admits that he is intimately close to Christianity: Christianity lives in and through him. Rather, I argue that it is by acknowledging his debt to Christianity and learning to love it as the condition enabling his *Zarathustra* that he can overcome the life-denying effects of Christianity still at work in his own relationship to himself, to others, and to life.

5. Scholars sometimes link Nietzsche's sickness and his philosophy by crediting the bursts of euphoria in his writing to experiences of mental illness, or by attributing his aphoristic style to the time he had to write between migraines. I am suggesting a complementary interpretation. Of course, Nietzsche eventually succumbed to his disease ten years before he died, losing his ability to write. However, achieving "great health" does not mean that one will never be ill and never die. Rather, "great health" represents an ability to learn from, grow through, and create out of whatever happens. This process, moreover, is not equivalent to positive thinking or having a good mental attitude. It involves disciplining ourselves to our sensory immersion in the moment and exercising our bodily becoming, such that we are able to tap energies of our own creative vitality and from this perspective, transform our experience of suffering from an occasion that depresses and horrifies to one that deepens our love and gives us more life.

Part 2   Isadora Duncan

1. See Daly (1995, 62–4) for an extended description of Duncan's performance aesthetic. For an example, see Carl van Vechten's review of Duncan's 1911 New York performance (Padgette 1974: 18–9). Van Vechten recalls green curtains and rose-colored light.

4   A Dionysian Artist

1. Biographical information for Duncan is drawn from biographies by Ann Daly, *Done Into Dance* (1995); Frederika Blair, *Isadora: Portrait of the Artist as a Woman* (1986), and Victor Seroff, *The Real Isadora* (1971), as well as from Duncan's autobiography, *My Life* (1928, hereafter "ML"), and two collections of her essays, *Art of the Dance* (1928, hereafter "AD") and *Isadora Speaks* (Ed. Franklin Rosemont. 1981, hereafter "IS"). See also Daly, "Isadora Duncan's Dance Theory," in *Dance Research Journal* (26[2], Fall 1994), and the chronology provided by Lillian Loewenthal in *The Search for Isadora* (1993).

2. While most commentators on Duncan note that she was influenced by Nietzsche's writings, few investigate this relationship with any depth. When they do, Nietzsche's work is often mischaracterized. For one example, "Like Nietzsche who formulated a philosophy of art derived from the two deities,

Apollo and Dionysus, Isadora celebrated the rapturous, exalted emotion and vital life essence of her dance as Dionysian" (Loewenthal 1993:24). While the Dionysus/Apollo pair was significant for Duncan, it was so in the context of Nietzsche's analysis of *religion*. There is no mention made of Duncan's investment in Nietzsche's larger philosophical project, the revaluation of all values.

3. In the translation Duncan uses, the passage reads: "If my virtue be a dancer's virtue, and if I have often sprung with both feet into golden-emerald rapture, and if it be my Alpha and Omega that every thing heavy shall become light, every body a dancer and every spirit a bird; verily, that is my Alpha and Omega" (ML Frontispiece). See discussion in chapter 2.

4. See also in *Art of the Dance*, "this vision is not impossible to realize, for I have seen the little children of my school, under the spell of music, drop all materiality and move with a beauty so pure that they attained the highest expression of human living" (AD 119). Daly notes that Duncan saw herself as "a goddess, or prophet, like Nietzsche's Zarathustra." But Daly mischaracterizes Zarathustra's vision as "a gospel of beauty built from love and art" (10), a quest for a spiritual rather than a physical–spiritual perfection. Compare with the reading of *Zarathustra* in chapter 2.

5. For further discussion of these issues in relation to contemporary philosophy, see LaMothe (2005a).

6. Jowitt (1988) notes how Duncan "cheerfully ignored" Nietzsche's opinion of women as "second class citizens" (94).

7. For accounts of dance and Christianity in the American context, see *Chronicles of the American Dance: From the Shakers to Martha Graham*, ed. Paul Magriel (1978). In support of Duncan's perceptions of "Puritan" America, see Ann Wagner, *Adversaries of Dance: From the Puritans to the Present* (1997).

8. Richard Schact describes these two texts as exercises in aesthetic education. See his essay, "Zarathustra/*Zarathustra* as Educator," in *Nietzsche: A Critical Reader* (1995).

9. Banes (1998) argues that Duncan's "calls for women's emancipation were tempered by her elision of women's identity with nurture and nature" (75). However, by reading Duncan's relation to "nature" in the context of her relationship to the Greeks and to Nietzsche, I argue that the way in which Duncan appealed to nature was itself liberating in the ways it encouraged and empowered women to be sites of their own nature-creation.

10. Daly (1995) contends that Duncan's appeals to the natural body and to Nietzsche served to produce and reproduce "social divisions along the lines of class and races" (113) as a way of distinguishing her dance and its audience from all others. By reading the natural body as dynamic, my reading offers an alternative to Daly's, highlighting the radical political implications of Duncan's claim.

11. Compare with Martin (1947) when he writes that "soul" meant "simply that correlative of the mind which produced, instead of intellectual concepts, quite irrational expressions of feeling . . . She had, however crudely and in whatever inaccurate and unscientific terminology, discovered the soul to be what less imaginative men called the autonomic system" (5).

12. In citing this account from Duncan's autobiography, I do not assume that the event actually happened as rendered; I read her description as evidence of what the event meant to her.

13. Daly (1995:144–54) describes the evolution of this idea in Duncan's mind and discusses particular tragedies she danced.
14. In this way Duncan's claim to be the chorus counters those who portray her work as "utopian," or promoting an ideal world of edgeless harmony. For example, see Graff (1997:17, 21, 50).

5   Incarnating Faith

1. See Franko (1995) who identifies "spontaneity" as the guiding principle of Duncan's theory of expression and claims that her career was "muddled by reputed paganism" (26). Or as one Duncan biographer explains: "it is clear that all consider the technique of Isadora negligible but that most view her principles and the impact of her career as enduring" (Terry 1963:166). The reasons for this failure, according to historians, range from psychological to circumstantial. Summarizing a widely held view, Ted Shawn states: "Isadora and the Greek Ideal freed us in the first two decades of this century from Victorianism and Puritanism. But in my opinion, she went much too far in the freedom—or license—she permitted herself" (Terry 1963:165).
2. A comparison with Nietzsche on this point is revealing. The debate over whether his writing is "philosophy" or "literature" raises the question of his *technique*, and whether he has any; the debate over whether or not he is a nihilist or relativist, raises the question, needless to say, of his *morality*. Yet in the case of Nietzsche, it is easier to see what these criticisms express: the challenge of evaluating one moral paradigm through the lens of another. Through the lens of Christian values both Nietzsche and Duncan can appear as disrespectful of classical traditions, and irresponsible in their personal creations.
3. Although I do not discuss it here, music was especially important for Duncan in inspiring both her practice and choreography. She and her students danced to Wagner, Gluck, Bach, Scriabin, Chopin, Schubert, Liszt, and Beethoven, among others.
4. On stage Duncan came alive. Commentators agreed that she had a charisma whose impact transcended the particulars of her form and dances. "The Holy Isadora" they called her in Berlin (Loewenthal 1993:193). See also Duncan's description of actress Eleanor Duse (AD 121–2).
5. For a discussion of the debates over Duncan's feminism, see LaMothe (2003).

Part 3   Martha Graham

6   An Affirmation of Life

1. For further discussion of this dance, see Banes (1998:74–80), and LaMothe (2003:351–73).

2. For biographical information on Martha Graham, I rely primarily on Agnes de Mille, *Martha* (1991a); Ernestine Stodelle, *Deep Song: The Dance Story of Martha Graham* (1984); Don McDonaugh, *Martha Graham* (1973); Deborah Jowitt, *Time and the Dancing Image* (1988); Elizabeth Kendall, *Where She Danced* (1979), as well as on Graham's essays, published *Notebooks* (1973), and autobiography, *Blood Memory* (1991).

3. Dance scholars highlight different pivotal moments in this interval. In 1927, for example, Duncan died and Graham gave her first concert of original compositions. In January 1930, Graham, Helen Tamiris, Doris Humphrey, and Charles Weidman combined forces to rent a theater on 39th Street in New York City, performing on alternating nights. In the words of John Martin, first dance critic appointed at the *New York Times*, "The American dance has come of age" (New York Times, January 5, 1930). See Julia Foulkes, *Modern Bodies* (2002), Introduction.

4. For a classic expression of this reading, see Walter Sorrell:

> If Isadora was a beginning, Martha was its fulfillment. In retrospect, it seems that Isadora's method was, as a most personal expression, limited to simple walking, running and leaping. It was soul minus technique, while Martha is technique plus soul. What was still a never-ending groping for new form and content with Isadora—vague, however impressive it may have been—has found in Martha its master. (Sorrell 1951:34–5)

5. Franko (1995) critically describes this approach: "Modernist accounts of modern dance history thus perform the telos of aesthetic modernism itself: a continuous reduction to essentials culminating in irreducible 'qualities' " (ix). The individualistic, modernist approach to dance history has given way to a plethora of projects seeking to contextualize the work of the "masters" in relation to movements in modern art, social issues, political concerns, European culture, and other forms of social and popular dancing. Franko, for example, sets canonical figures in context and plots "internal critiques of expression theory" as driving the relations among them. Yet the "rejection of subjectivism" he notes in the relation of Graham to Duncan still honors the narrative in which the religious language and thematic materials of the early modern dancers appear as accidental to the work of creating good dance (xii, chapters one and three). I argue that Graham did not reject Duncan's religious claims, but rather approached their realization differently.

6. Those succeeding generations of dancers who countered Graham by arguing that movement always lies did not refute her meaning: whether you are lying or not, the movement reveals it. See for example Paul Taylor, *Private Domain* (1987).

7. McDonaugh is one commentator who notes this paradox: "The Puritan energy and vigor of America drove and shaped all of Graham's work, but did not blunt her sense of inquiry. That same energy drove Graham herself and shaped her day" (McDonaugh 1973:68).

8. For further discussion, see LaMothe, "Passionate Madonna" (1998). In the years after Graham left the company, St. Denis developed the position that Christian religion must be danced in order to release its wisdom into the world.

9. Nietzsche's writings seem to have exerted the greatest influence on Graham in the 1930s, as she was crystallizing her vision for dance and discovering the generative kernel of her dance practice (see chapter 7). Graham turned down an invitation to perform in the 1936 Berlin Olympic Games due to her condemnation of the Nazi regime. As is known, the Nazis also (mis)read Nietzsche, harnessing his philosophy to the anti-Semitic projects he abhorred actively during his lifetime. This confluence of events may have contributed to Graham's resistance to reference his works and name directly after this time. His influence, however, endured in her vision for dance and the dance process she evolved to realize it. See Manning (1993:255–85) for a discussion of Graham in relation to German nationalism.

10. Graham choreographed other dances during this time with Nietzsche-related themes, including: *Dithyrambic* (1931), *Bacchanale* (1931), *Ecstatic Dance* (1932), *Bacchanale No. 2* (1932), *Choric Dance for an Antique Greek Tragedy* (1932), and others. These dances appeared in strong contrast to her Denishawn-inspired dances of 1926–1928. See de Mille (1991a:434–55) for one of the more complete lists of Graham's dances.

11. Graff (1997:chapter 5) for example, writes: "In Graham choreography, audiences identified with the individual, not the mass, which amounted to a kind of heresy within the workers' dance movement" (107). Graff is evaluating Graham's politics. Such dances demonstrate to Graff Graham's "political aloofness" (108): she was not interested in "group efforts and revolutionary dance movements" (107). The interpretation offered here attributes the dance's success to the fact that audience members identify with both the individual and the group.

   One scholar who notes religious implications is Lloyd (1949): "It is the essence of the eternal struggle of the individual with something new to offer, coming up against the blank wall of conservatism in any field—religion, art, science, or private life" (51).

12. See also: "We look to the dance to impart the sensation of living in an affirmation of life, to energize the spectator into keener awareness of the vigor, the mystery, the humor, the variety, and the wonder of life" (G35, 58); and "There must be something that needs to be danced. Dance . . . is not an emotional catharsis for the hysterical, frustrated, fearful or morbid. It is an act of affirmation, not of escape" (G41, 184); or "Falls are used primarily as preliminary to and therefore as a means of 'affirmation.' In no fall does the body remain on the floor, but assumes an upright position as part of the exercise. My dancers fall *so they may rise*" (G41, 187). The examples go on.

13. See also an essay written for *Revolt of the Arts* (1930) where Graham writes: "Like the modern painters and architects, we have stripped our medium of decorative unessentials. Just as fancy trimmings are no longer seen on buildings, so dancing is no longer padded. It is not 'pretty' but it is much more real" (Armitage 1937:97). In vowing to "strip away" what is *merely* "decorative," Graham embraced the modern notion that the process of stripping or abstraction leads one to the source of significance. She would "abstract" the "essence" of dance, in the sense of squeezing an orange for its juice and discarding the pulp (Lloyd 1949:52). The debate on whether Graham and other modern

dancers in the 1920s and 1930s were sufficiently "modern" continues. See Burt (1998:13–15, 123–6, and 160–3); Franko (1995:chapter 3, 38–74); Banes (1987), and Manning (1993). The issues concern: was her movement *abstract* (or emotional)? Was it *reflexive* (as Clement Greenberg requires, or not)? Was its engagement of the *primitive* in line with a modernist aesthetic (or not)? I respond "yes" to all three. I agree with Burt that Graham is a "modernist" when "modernism" is defined as "a progressive deconstruction of outmoded aesthetic conventions and traditions" (15).

14. Of course, in order for dance to communicate participation in an experience, the spectators themselves must be willing to open to receive the dance—open enough to allow their instincts for gestural imitation and visceral identification to come into play. Graham admits as much. A dance is only as good as the audience allows it to be. See discussion in chapter 8.

15. Franko (2002:66–71) discusses the influence on Graham's theories of rhythm in Mary Austin's book *American Rhythm* (1930).

16. This approach to Graham's "American" themes bypasses the questions of which dances were "American" and of how to explain her "American" period. From this perspective here, all of her dances were "of America"—and in part due to Nietzsche's influence—whether or not the themes were explicitly American because of the way she engaged the materials. Dance with explicitly American themes do not represent a departure, nor a new attempt to be more theatrical or accessible. They represent an extension of what she was already doing. For a discussion of Graham as practicing "isolationism," see Burt (1998:chapter 7). In support of his position, Burt even cites her Nietzschean phrase: "One must become what one is" (136).

17. For a discussion of primitivism in modern dance, see Burt (1998:chapter 8); Franko (1995:chapter 3). Burt claims that Graham was operating with a "common-sense" notion of a distinction between primitive and western societies (160). He acknowledges that Graham uses the term to stress the similarity of the "primitive" and the "modern" experience, but then criticizes her for giving others an opportunity to talk about the "primitive" as childlike (164).

18. Other dances as well enact this confluence of themes, including *Celebration* (1936), *Two Primitive Canticles* (1931), and *El Penitente* (1940).

19. See Burt (1998:181–9); de Mille (1991:177–83), for a discussion of the dance and its reception. De Mille writes: "dancing as a medium had taken a step into new and hitherto forbidden realms. The only art to have been separated from religion, dance had shriveled and starved since 1295, when the Christian church by formal edict proscribed it. Martha threw open the great door" (182). While dancing did occur in Christian contexts after 1295, the spirit of de Mille's statement reveals the sense of significance the dance represented for those involved.

20. As Burt (1998) observes: "The Virgin does very little dancing in *Primitive Mysteries* and is often still while the others dance, yet she always remains the center of attention . . . it is only the Virgin who seems, at the two moments when she makes the sign of the cross, to experience any connection between individual and a deity" (187). I would disagree that only the Virgin experiences such connection. The others experience it in their dancing and in relation to Her, which is part of

what the dancing communicates. The dance does not communicate the ability of movement "to lead to disassociational states of consciousness" per se (187), but the ability, as Burt says of Graham's relation to the "primitive," to enact a permeable relation between self and Other. Where Burt writes that Graham's dance "articulates a conflictual fear and fascination with the Other that itself articulates in sublimated form a deep ambivalence about modernity itself" (189), I argue that Graham's dance frames dance as a resource for *overcoming* the physiological contradiction of "modernity" by transforming our experience of that contradiction, and calling us to acknowledge our responsibility in the creation of the "Other."

21. Anna Sokolow notes that it was the experience of performing this dance that "led me toward what I wanted to do—create religious dances. I left the Graham company to do just that, and Martha showed me the path I needed to follow" (Horosko 1991:45).

7   Athletes of God

1. See also Stodelle (1984) 48, 57.
2. "St. Denis' . . . yoga classes may have provided inspiration for Graham's floorwork, and her interest in breathing may have led to Graham's development of the contraction and release principle, as Gertrude Shurr and others have suggested" (Helpern 1994:7).
3. See *Theater and Its Double* (1958).
4. Dorothy Bird Villiard, "Martha Graham and her Early Technique," Lecture at Adelphi University, New York Public Library, Dance Collection.
5. For a discussion of dance as enacting the possibility of human freedom, see Sondra Horton Fraleigh, *Dance and the Lived Body* (1987).
6. See also Graham's comment: "I believe that dancing can bring liberation to many because it brings organized activity. I believe that the exercises I use are as right for a lay person as for a professional dancer, because they do no violence, anatomically or emotionally . . . I have always thought first of the dancer as a human being" (G41, 184). Of course, she was later criticized for developing a technique based on her own body that caused violence to the anatomies of others.
7. For further evidence of this point, see Graham's oft-repeated phrase: "It takes ten years to build a dancer. The body must be tempered by hard, definite technique—the science of dance movement—and the mind enriched by experience. Both must be ripened to maturity before the dancer has a message he can convey effectually. A part of this maturity is muscular memory. It has nothing to do with so-called natural dancing or improvisation" (G37, 106). Not only can a dancer progress only as quickly as her body will allow, that physical progression must happen in concert with an enriching of the mind "by experience."
8. In the years after World War II, many of Graham's dancers received ballet training, and she began to use this training in her dances for its movement possibilities. As Graham dancer May O'Donnell reports, there was a greater emphasis

on physical speed, precise lifts, and greater overall virtuosity during this time (Horosko 1991:70).

9. See the opening pages of *Blood Memory* (1991) for these references and more. As Helpern notes: "Dance, life, religion all intermingle in Graham's aesthetic" (1994:25). This analysis interprets the significance of that "intermingling" for her dance process and philosophy.

10. Note as well that in many Christian traditions, *spirit* is associated with *breathing* as a metaphor for divinity. Spirit, the breath of life, moves across the deep before the moment of creation; spirit moves tongues to speak at Pentecost. Thus, in choosing to ground her technique in the movements of breathing, Graham locates her practice at the core of the western religious imagination as a primary expression of *spirit*.

11. "Martha Graham and Early Technique," Lecture at Adelphi University, New York Public Library.

12. Graham makes this comment when reflecting upon an experience she had while performing *Deaths and Entrances*, her dance based loosely on the three Bronte sisters. In this experience she "suddenly knew" that "witchcraft . . . is the being within each of us . . . of creative energy" (N 85).

13. Straus (1966:147). See also: "Upright posture pre-establishes a definite attitude toward the world; it is a specific mode of being-in-the-world," Straus (1966:139).

14. De Mille (1991:98). See also Helpern (1994:23).

15. Mircea Eliade, *The Sacred and the Profane* (1987).

16. Graham also talked about the spiraling path of energy around the spine in terms of "kundalini." Kundalini, meaning the "coiled-up one," symbolizes energy coiled at the base of the spine that practitioners of kundalini yoga seek to awaken and move upward through the seven chakras. McDonaugh recounts how Graham walked into a class being taught by her assistant, Jane Dudley, and took over because the class lacked the right "feel"—"There was no 'kundalini' " McDonaugh (1973:196). See also (B 122).

17. Compare with someone who interprets interdependence as a unity: "Graham recognized the interdependence of physical and emotional elements in the movements of the torso and the pelvis, and she centered her technique on the enhancement of that interdependence. In this she proved so successful that dancers cannot fail to realize, in the routine activity of daily class, that the physical and the emotional are one" (Helpern 1994:24).

18. I argue that it is practices of reading and writing that help us sustain an image of ourselves as having inner and outer selves. Most Graham commentary stumbles on this distinction, insisting that Graham's work succeeds in "expressing personal and collective inner worlds . . . making tangible the otherwise hidden forces that drive the human being" (Horosko 1991:4). It is this quality of her dance that people often point to as its link to religion: "It is akin to religion in that it gives concrete form to the unseeable spirit of mankind, the hidden essence of human life" (Horosko 1991:4). Such readings miss the Nietzschean critique of the distinction between soul and body that Graham picks up from him and from Duncan. As argued here, Graham enacts the dynamic creativity

of bodily becoming in making its own inner life. In Graham's words: "In dance, each time, it may be a mystical or religious connotation that you feel, but principally it is the body exalting in its strength and its own power" (Horosko 1991:145). Graham's point is not to reduce the "religious connotation" to what is material, but to acknowledge the creative role of our exalting bodies in our sense of what counts as "mystical or religious" meaning.

8   Words to Dance

1. The music was composed by Halim El-Dabh; the set created by Isamu Noguchi, and the lighting designed by Jean Rosenthal. Graham and Helen McGehee designed the costumes (de Mille 1991a:449).
2. For further discussion about the relationship between text and dance in St. Denis' work, see LaMothe (1998), "Passionate Madonna." St. Denis spoke and wrote beautifully about how people must dance in order to release the wisdom of Christian texts into the world. She argued that dance was necessary given its enabling relationship to textual traditions. Yet she assumed a relationship of complementarity that in the end fails to challenge the privilege accorded to texts over and against dancing in western cultures. Towards the end of her life, her goal was to discover a gestural language for the west on par with the system of Indian mudras used in the forms of Indian classical dance.
3. Franko (1995) for example, in interpreting the significance of this dance, argues that Graham here moves from modernism "into the emotivist mainstream" (68). After this point her work, Franko writes, became "operatic in the Nietzschean sense, when an expressive veneer became applied to [her work's] formulaic vocabulary" (51). His reading could benefit from illuminating the Nietzschean connection and context within which this dance appears. See also Graff (1997:124–31) who claims that this acclaimed, politically relevant dance incorporates many elements that had "proved popular in the revolutionary dance movement, including moral fervor, archetypal figures, pageantry, and text" (127). As Graff writes, "It had been done before . . . but not by Graham" (129). See also Helpern (1994:16), and Foulkes (2002:147–52).
4. Dorothy Bird, for example, explains: "Martha said, 'Dance has nothing to do with what you can tell in words. It has to do with actions, colored by deep inarticulate feelings that can only be expressed in movement.' She did not permit a single sentence, neither a subject nor an object to be considered as a basis for movement, only verbs and adverbs. Those were the only words she had in her notebook at the time" (Horosko 1991:48).
5. For a discussion of this phrase as well as van der Leeuw's theories of dance and religion, see LaMothe (2004:chapters 5–9).
6. See Stodelle (1984:105–7); de Mille (1991:232–6), and Soares (1992:40–1).
7. Manning and others point out that the bodies allowed to represent the diversity of American bodies were themselves all white. See Manning (1996),

"American Document and American Minstrelsy" in G. Morris (ed.) *Moving Words: Re-Writing Dance.* See also Burt (1998:130–1).

8. Jane Dudley, in a discussion of Graham's *Letter to the World,* inspired by the poetry of Emily Dickenson, confirms: "Martha really felt that Puritanism was the death element in life, and sexuality was the life element in life. That conflict is very important in her work" (17). I argue that Graham's dances enact a more complex tension in which our relationships to "Puritanism". . . and "sexuality" are both experienced as physiological contradictions. As discussed in relation to *American Document,* Graham reveals the sexuality in Puritan symbols and the puritanical elements in our desire for sexual intimacy. See "Martha Graham's Early Technique and Dance: The 1930s, A Panel Discussion" (Helpern 1999:7–31).

9. Banes (1998) elucidates the specifically sexual dimension of the physiological contradiction Graham dances in her analysis of *Night Journey* (1947). "Several commentators have observed that Graham's choreography above all disclosed the conflict between sexual pleasure and puritanical repression. Nowhere in Graham's *oeuvre* is this psychological and cultural warfare more poignantly articulated . . . [Jocasta] is the extreme metaphor for the profound ambivalence that women in Graham dances generally feel toward their sexual desires" (163). Banes suggests biographical and cultural reasons (164–7) and concludes that her dance takes one step forward and two backwards, "undermining its sexually liberatory aspect with its gender hierarchy and its emotional extremes . . . Jocasta's erotic enthusiasm was a tragic mistake, for it only signaled her doom" (167). The Nietzschean perspective provided here, as exemplified in the analysis of *Clytemnestra* later in this chapter, offers an alternate interpretation, one that identifies transforming and affirmative affects on our (Puritan-informed) experience of sexuality of *dancing* a tragedy.

10. See Nolini Barretto, "The Role of Martha Graham's Notebooks in Her Creative Process," in Helpern (1999:53–67). Barretto describes the *Notebooks* as a "springboard from which she launches into the next phase of the creation of the dance," namely writing a script for the composer (64–5). She notes that few critics have attempted an in-depth study of the *Notebooks,* and that those who have have not found that they added much to our understanding of the dances described (66). *Clytemnestra* at least is an exception in this regard.

11. The earliest dance for which there are notes is her 1943 *Deaths and Entrances,* a dance inspired by the writing of the Bronte sisters. Yet the notes under this dance represent her reflections on a performance dated February 1950, during which she came to a deeper understanding of the dance's meaning. The next earliest dances include *Appalachian Spring* in 1944 and *Dark Meadow* in 1946.

12. Graham's religious references are too extensive to document here. She quotes from the Bible a number of times (Exodus, 2 Samuel, Matthew, Luke, John, Ephesians, Revelation, and others) as well as from Christian apocryphal literature (Acts of John, Tobit, Tobias). She cites scholars of religion (Joseph Cambell, Evelyn Underhill, Mircea Eliade, A.K. Coomaraswamy, F.M. Cornford, Carl Jung, Sigmund Freud, Santayana, Schopenhauer, and others). She cites poets and writers who engage religious themes and materials

(T.S. Eliot, Emily Dickinson, Andre Gide, Plato, Coleridge, Basho, Rilke, Novalis, Holderin, William Blake, Sartre, and others). References to Nietzsche are conspicuously absent.

13. See, for example, in relation to Joan of Arc: "If the story of Joan of Arc is to be danced then its substance concerns that which cannot be spoken—It is a conflict each knows in lesser or greater degree . . . / There is a constant conflict between / common sense / comfort / and / that which must be done—/ The vision" (N 322).

14. Anna Kisselgoff, *New York Times*, March 27, 1973.

15. For further discussions of *Clytemnestra*, see Helen McGehee, "An Opportunity Lost" in Helpern (1999:69–76). McGehee describes this dance as "the culminating achievement of this country's preeminent genius, the crown of her illustrious career, both as a dancer and, more importantly, as a choreographer" (69).

16. Taped interview with David Raher (September 12, 1973). New York Public Library, Dance Collection.

17. Score, Vol I, 42.

18. Graham is recognized as pioneering this "flashback" technique which is now common in films, for example. In Graham's hands, this technique is not used to tell a story whose end is known, but to put on display the pressures driving a person to act in ways that realize her "self." Graham adopts this technique at the beginning of her "Greek cycle" with *Cave of the Heart* in 1946. *Cave* was based on the story of Medea. *Errand into the Maze* (1947) and *Night Journey* (1948) were inspired by the stories of Theseus/Ariadne and Jocasta respectively. After *Clytemnestra*, she was to dance *Alcestis* (1960) and *Phaedra* (1962).

19. Score, Vol II, 49.

20. See Walter Terry: "[It is] not a dance duplication of either the myths or the ancient dramas," (*New York Herald*, April 2, 1958); and "it is in no sense a danced version of these legends. Rather it is an emotional communication and revelation developed in ritualistic form," (*Dance and Dancers*, 6–58, 18–19). *Clytemnestra*, Clippings File, New York Public Library, Dance Collection.

21. *Errand into the Maze* was first performed on February 27, 1947 at the Ziegfeld Theater in New York City. The music was composed by Gian Carlo Menotti; the decor by Isamu Noguchi, and the lighting designed by Jean Rosenthal. Graham designed the costumes (de Mille 1991:447). This dance appears in *Martha Graham: Three Contemporary Classics*, a program first aired on *Dance in America* in 1984 as "An Evening of Dance and Conversation with Martha Graham." The videotape includes *Errand into the Maze*, *Cave of the Heart*, and *Acts of Light*.

22. The first performance of *Acts of Light* was on February 26, 1981 at City Center in New York City. The music was by Carl Nielson; costumes by Halston; and lighting by Beverly Emmons (de Mille 1991:454).

23. Program ca. 1954 from tour in Japan. New York Public Library, Dance Collection.

24. Even Foster (1986) who describes Graham's movements as erupting "out of the tension between inner self and outer body" (25) reifies the distinction between inner and outer as something that the movement crosses rather than a distinction the movement itself serves to create. As such, while she does an excellent

job of analyzing how various practices of dance serve to make a dancing body, there is more to be done in understanding the creative agency of "the body" in that process. See chapter 1.

25. B 237. See also the videotape *Martha Graham: Three Contemporary Classics* (1984), Introduction to *"Acts of Light."*

26. Compare this interpretation with Foster (1986) who interprets the sentence to mean that movement "signals a person's true identity and feelings" (28). Foster's reading abides by the logic of the ascetic ideal.

Conclusion: Another Ideal

1. A description of performing *On Fire*, conceived, choreographed, and performed by LaMothe under the auspices of the Religion and Art Initiative of the Center for the Study of World Religions, at the Rieman Center for the Performing Arts, Harvard University, May 2004.

Bibliography

Friedrich Nietzsche (1844–1900)

Nietzsche, Friedrich. 1914. "On Truth and Falsity in their Ultramoral Sense," in *The Complete Works of Friedrich Nietzsche*. Ed. Oscar Levy. Volume Two. New York: Macmillan.

———. 1954. *The Portable Nietzsche*. Ed. Walter Kaufmann. New York: Penguin.

———. 1966. *Beyond Good and Evil*. tr. Walter Kaufmann. New York: Random House.

———. 1967, *The Birth of Tragedy and The Case of Wagner*. Ed. and tr. by Walter Kaufmann. New York: Vintage Press.

———. 1974. *The Gay Science with a Prelude in Rhymes and an Appendix of Songs*. New York: Vintage Press.

———. 1980. *Kritische Studienausgabe in 15 Bänden*. Ed. Giorgio Colli and Mazzino Montinari. Munich: Deutscher Taschenbuch Verlag; Berlin: Walter de Gruyer.

———. 1983. *Untimely Meditations*. tr. R.J. Hollingdale. Cambridge: Cambridge University Press.

———. 1984. *Human, All Too Human: A Book for Free Spirits*. tr. Marion Faber with Stephen Lehmann. Lincoln and London: University of Nebraska Press.

———. 1989. *On the Genealogy of Morals and Ecce Homo*. Ed. and tr. by Walter Kaufmann. New York: Vintage Press.

Nietzsche: Secondary Bibliography

Atwell, John E. 1984. "The Significance of Dance in Nietzsche's Thought," in *Illuminating Dance: Philosophical Explorations*. Ed. Maxine Sheets-Johnstone. Cranbury, NJ: Associated University Press.

Bergoffen, Debra. 1990. "Posthumous Popularity: Reading, Privileging, Politicizing Nietzsche," *Soundings* 73, I (Spring): pp. 37–60.

Blondel, Eric. 1991 [1986]. *Nietzsche: The Body and Culture*. tr. Sean Hand. Stanford: Stanford University Press.

Bloom, Allen. 1987. *The Closing of the American Mind: How Higher Education has Failed Democracy and Impoverished the Souls of Today's Students.* New York: Simon & Schuster.

Burgard, Peter J., Ed. 1994. *Nietzsche and the Feminine.* Charlottesville: University of VA Press.

Clark, Maudemarie. 1990. *Nietzsche on Truth and Philosophy.* Cambridge and New York: Cambridge University Press.

Danto, Arthur. 1965. *Nietzsche as Philosopher.* New York: Macmillan.

Deleuze, Gilles. 1983. *Nietzsche and Philosophy.* tr. Hugh Tomlinson. New York: Columbia University Press.

Derrida, Jacques. 1979 [1978]. *Spurs: Nietzsche's Styles/Eperons: Les Styles de Nietzsche.* tr. Barbara Harlow. Chicago: University of Chicago Press.

———. 1985. *The Ear of the Other: Otobiography, Transference, Translation.* Ed. Christie McDonald. tr. Peggy Kamuf for French edition. Eds. Claude Levesque and Christie McDonald. Lincoln & London: University of Nebraska Press.

Eagleton, Terry. 1990. *The Ideology of the Aesthetic.* London: Blackwell.

Gillespie, Michael Allan and Tracy Strong, eds. 1988. *Nietzsche's New Seas: Explorations in Philosophy, Aesthetics, and Politics.* Chicago: University of Chicago Press.

Gooding-Williams, Robert. 2001. *Zarathustra's Dionysian Modernism.* Stanford: Stanford University Press.

Graybeal, Jean. 1990. *Language and the "Feminine" in Nietzsche and Heidegger.* Bloomington: Indiana University Press.

Hayman, Ronald. 1980. *Nietzsche: A Critical Life.* New York: Oxford University Press.

Heidegger, Martin. 1991 [1961]. *Nietzsche. Volumes One and Two.* tr. David Farrell Krell. Harper San Francisco.

Higgins, Kathleen. 1987. *Nietzsche's Zarathustra.* Philadelphia: Temple University Press.

Irigaray, Luce. 1991 (1980). *Marine Lover of Friedrich Nietzsche.* tr. G.C. Gill. New York: Columbia University Press; *Amante Marine: De Friedrich Nietzsche.* Paris, Minuit.

———. 1993 (1987). *Sexes and Genealogies.* tr. Gillian C. Gill. New York: Columbia University Press.

Kaufmann, Walter. 1974 (1950). *Nietzsche. Philosopher, Psychologist, Antichrist.* Princeton: Princeton University Press.

Koelb, Clayton, Ed. 1990. *Nietzsche as Postmodernist: Essays Pro and Contra.* Albany, NY: State University of New York.

Kofman, Sarah. 1972. *Nietzsche et la Metaphore.* Paris: Payot; 1993. tr. Duncan Large. London: Athlone Press.

Krell, David Farrell. 1986. *Postponements: Women, Sensuality and Death in Nietzsche.* Bloomington: Indiana University Press.

Krell, David Farrell and David Wood, eds. 1988. *Exceedingly Nietzsche: Aspects of Contemporary Nietzsche Interpretation.* London: Routledge.

Magnus, Bernd and Kathleen Higgins. 1996. *The Cambridge Companion to Nietzsche.* Cambridge: Cambridge University Press.

Murphy, Tim. 2001. *Nietzsche, Metaphor, Religion.* Albany: State University of New York Press.

Nehamas, Alexander. 1985. *Nietzsche: Life as Literature.* Cambridge: Harvard University Press.

Oliver, Kelly. 1993. "A Dagger Through the Heart: Reading Nietzsche's On the Genealogy of Morals." *International Studies in Philosophy.* XXV: 2.

———. 1995. *Womanizing Nietzsche: Philosophy's Relationship to the "Feminine."* New York: Routledge.

Oliver, Kelly and Marilyn Pearsall. 1998. *Feminist Interpretations of Friedrich Nietzsche.* University Park: Pennsylvania State University Press.

O'Flaherty, James C., Timothy F. Sellner, and Robert M. Helm, eds. 1985. *Studies in Nietzsche and the Judeo-Christian Tradition.* Chapel Hill: University of North Carolina Press.

Peters, Heinz Friedrick. 1985. *Zarathustra's Sister: The Case of Elisabeth and Friedrich Nietzsche.* Marcus Wiener.

Richardson, John. 1996. *Nietzsche's System.* New York: Oxford University Press.

Roberts, Tyler. 1998. *Contesting Spirit: Nietzsche, Affirmation, Religion.* Princeton: Princeton University Press.

Santaniello, Weaver, Ed. 2001. *Nietzsche and the Gods.* Albany: State University of New York Press.

Schact, Richard. 1983. *Nietzsche.* London: Routledge and Kegan Paul.

Schrift, Alan. 1990. *Nietszche and the Question of Interpretation: Between Hermeneutics and Deconstruction.* New York: Routledge.

Solomon, Robert C. 1980. *Nietzsche: A Collection of Critical Essays.* Notre Dame: University of Notre Dame Press.

Art, Religion, and Philosophy

Alcoff, Linda. 1997. "Cultural Feminism versus Post-Structuralism: The Identity Crisis in Feminist Theory," in *The Second Wave.* Ed. Linda Nicholson. London & New York: Routledge, 330–55.

Artaud, Antoinin. 1958 [1938]. *The Theater and its Double.* tr. Mary Caroline Richards. New York: Grove Weidenfeld.

Caputo, John. 2000. *More Radical Hermeneutics.* Bloomington: Indiana University Press.

Christ, Carol. 2003. *She Who Changes: Re-Imagining the Divine in the World.* New York: Palgrave Macmillan.

Clark, Suzanne. 1991. *Sentimental Modernism: Women Writers and the Revolution of the Word.* Bloomington: Indiana University Press.

Derrida, Jacques 1976 [1967]. *Of Grammatology.* tr. Gayatri Spivak. Baltimore: The Johns Hopkins Press.

———. 1981 [1972]. *Dissemination.* tr. Barbara Johnson. University of Chicago Press.

———. 1982 [1972]. *Margins of Philosophy.* tr. Alan Bass. Chicago: University of Chicago Press.

Derrida, Jacques and Christie V. McDonald "Choreograpies," *Diacritics* 12(2): 66–76.

Eliade, Mircea. 1987. *The Sacred and the Profane.* Harcourt Brace.

Feuerbach, Ludwig. 1957. *The Essence of Christanity.* tr. George Eliot. New York: Harper Torchbooks.

Foucault, Michel. 1980. *Power/Knowledge: Selected Interviews and Other Writings 1972–1977.* Ed. and tr. Colin Gordon. New York: Random House, Pantheon Books.

Frankenberry, Nancy et al., eds. 1994. *Hypatia: Special Issue: The Feminist Philosophy of Religion.* 9(4).

Grosz, Elizabeth. 1989. *Sexual Subversions: Three French Feminists.* Australia: Allen & Unwin. Hollywood, Amy. 2002. *Sensible Ecstasy: Mysticism, Sexual Difference, and the Demands of History.* Chicago: University of Chicago Press.

———. 1995. *Space, time and Perversion: Essays on the Politics of Bodies.* New York: Routledge.

Irigaray, Luce. 1993. "Irigaray and the Divine," in *Transfigurations: Theology and The French Feminists.* Eds. C.W. Maggie, Kim, Susan M. St. Ville & Susan M. Simonaitis. Minneapolis, MN: Fortress Press.

———. 1993a [1987]. *Sexes and Genealogies [Sexes et Parentes].* tr. Gillian C. Gill. New York: Columbia University Press.

———. 1993b [1984]. *An Ethics of Sexual Difference.* tr. Carolyn Burke and Gillian C. Gill. New York: Cornell University Press.

Jantzen, Grace. 1999. *Becoming Divine: Toward a Feminist Philosophy of Religion.* Bloomington: Indiana University Press.

Jones, Serene. 1993. "This God Which is Not One: Irigaray and Barth on the Divine" in *Transfigurations: Theology and The French Feminists.* Eds. C.W. Maggie, Kim, Susan M. St. Ville & Susan M. Simonaitis. Minneapolis, MN: Fortress Press.

Kant, Immanuel. 1960. *Religion within the Limits of Reason Alone.* tr. Theodore Greene and Hoyt Hudson. New York: Harper Torchbooks.

———. 1987. *Critique of Judgement.* tr. W.S. Pluhar. Indianapolis: Hackett Publishing Company.

Kierkegaard, Soren. 1983. *Fear and Trembling and Repetition.* Ed. and tr. by Howard V. Hong and Edna H. Hong. Princeton: Princeton University Press.

Kristeva, Julia. 1980. *Desire in Language: A Semiotic Approach to Literature and Art.* tr. T. Gora, A. Jardine, L.S. Roudiez. New York: Columbia University Press.

———. 1982. *Powers of Horror: An Essay on Abjection.* tr. Leon S. Roudriez. New York: Columbia university Press.

LaMothe, Kimerer L. 1998. "Passionate Madonna: The Christian Turn of American Dancer Ruth St. Denis." *Journal for the American Academy of Religion* 66, 4 (Winter): 747–69.

———. 2003. "Giving Birth to a Dancing Star: Reading Friedrich Nietzsche's Maternal Rhetoric via Isadora Duncan's Dance." *Soundings* 86: 3–4.

———. 2004. *Between Dancing and Writing: The Practice of Religious Studies.* New York: Fordham University Press.

———. 2005a. " 'A God Dances Through Me': Isadora Duncan on Friedrich Nietzsche's Revaluation of Values." *Journal of Religion.* April. Vol. 85 No 2: pp. 241–66.

———. 2005b. "Reason, Religion, and Sexual Difference: Resources for a Feminist Philosophy of Religion in Hegel's *Phenomenology of Spirit." Hypatia.* Vol. 20, No. 1.

————. 2005c. "Why Dance?" *Method and Theory in the Study of Religion.* 17(2): pp. 101–33.

Langer, Susanne K. 1953. *Feeling and Form.* New York: Charles Scribner's Sons.

Lloyd, Genevieve. 1984. *The Man of Reason: "Male" and "Female" in Western Philosophy.* Minneapolis: University of Minnesota Press.

Miles, Margaret R. 1981. *Fullness of Life: Historical Foundations for a New Asceticism.* Philadelphia: The Westminster Press.

————. 1985. *Image as Insight: Visual Understanding in Western Christianity and Secular Culture.* Boston: Beacon Press.

————. 1991a. *Carnal Knowing: Female Nakedness and Religious Meaning in the Christian West.* New York: Vintage Books, Random House.

————. 1991b. *Desire and Delight: A New Reading of Augustine's Confessions.* New York: Crossroad.

————. 1996. *Seeing Is Believing: Religion and Values in the Movies.* Boston: Beacon Press.

————. 1998. "Image," in *Critical Terms for Religious Studies.* Ed. Mark C. Taylor. Chicago, IL: Chicago University Press.

Nicholson, Linda. 1999. *The Play of Reason: From the Modern to the Postmodern.* Ithaca, NY: Cornell University Press.

Parsons, Susan Frank, Ed. 2002. *The Cambridge Companion to Feminist Theology.* Cambridge: Cambridge University Press.

Schopenhauer, Arthur. 1995. *World as Will and Idea.* Abridged Edition, Ed. David Berman and tr. by Jill Berman. London: Everyman Paperback Classics.

Steiner, Wendy. 2001. *Venus in Exile: The Rejection of Beauty in Twentieth-Century Art.* New York: Free Press.

Straus, Edwin. 1966. *Phenomenal Psychology.* New York: Basic Books.

Taylor, Mark C. 1984. *Erring: A Postmodern A/theology.* Chicago: University of Chicago Press.

————. 1992. *Disfiguring: Art, Architecture, Religion.* Chicago: University of Chicago Press.

Torgovnick, Marianna. 1990. *Gone Primitive: Savage Intellects, Modern Lives.* Chicago: University of Chicago Press.

Van der Leeuw, Gerardus. 1948, 1955. *Wegen en grenzen. Studie over de verhouding van religie en kunst.* Amsterdam.

————. 1963. *Sacred and Profane Beauty: The Holy in Art.* tr. David Green. New York: Henry Holt.

Wollstonecraft, Mary. 1992 [1792]. *A Vindication of the Rights of Woman.* New York: Everyman's Library.

Dance Studies

Adams, Doug and Diana Apostolos-Ceppadona, eds. 1990. *Dance as Religious Studies.* New York: Crossroad.

Aeschylus. 1953. *Oresteia: Agamemnon, The Libation Bearers, The Eumenides.* tr. Richmond Lattimore. Chicago: University of Chicago Press.

Armitage, Merle, Ed. 1978 [1937]. *Martha Graham: The Early Years*. New York: de Capo Press, 1978.

Austin, Mary Hunter. 1930. *The American Rhythm: Studies and Reexpressions of Amerindian Songs*. Boston: Houghton Mifflin Company.

Banes, Sally 1987. *Terpsichore in Sneakers*. Wesleyan University Press.

———. 1993. *Greenwich Village 1963: Avant-garde Performance and the Effervescent Body*. Duke University Press.

———. 1998. *Dancing Women: Female Bodies on Stage*. London & New York: Routledge.

Collingwood, R.G. 1938. *The Principles of Art*. London: Oxford.

Blair, Frederika. 1986. *Isadora: Portrait of the Artist as a Woman*. New York: McGraw-Hill Book Company.

Burt, Ramsey. 1998. *Alien Bodies*. London & New York: Routledge.

Carter, Alexandra, Ed. 1998. *Routledge Dance Studies Reader*. London & New York: Routledge.

Collingwood, R.G. 1938. *The Principles of Art*. London: Oxford.

Copeland, Roger. 1990. "Founding Mothers: Duncan, Graham, Rainer, and Sexual Politics," *Dance Theatre Journal* 8, 3 (Fall).

Copeland, Roger and Marshal Cohen. 1983. *What Is Dance? Readings in Theory and Criticism*. New York: Oxford.

Cott, Nancy. 1977. *The Bonds of Womanhood: "Women's Sphere" in New England, 1780–1835*. New Haven: Yale University Press.

Daly Ann, 1994. "Isadora Duncan's Dance Theory." *Dance Research Journal* 26, 2 (Fall).

———. 1995. *Done Into Dance*. Bloomington, IN: Indiana University Press.

Davies, J.G. 1975. "Towards a Theology of the Dance," in *Worship and Dance*. Ed. J.G. Davies. University of Birmingham, ISWRA.

———. 1984. *Liturgical Dance: An Historical, Theological, and Practical Handbook*. London: SCM Press.

de Mille, Agnes. 1991a. *Martha: The Life and Work of Martha Graham*. New York: Random House.

———. 1991b. "Measuring the Steps of a Giant," *New York Times*, April 7.

Desmond, Jane C., Ed. 1997. *Meaning and Motion*. Durham, NC: Duke University Press.

Dudley, Jane. 1991. "Graham at Work," *Dance and Dancers*. May.

Duncan, Isadora. 1928a. *Art of the Dance*. New York: Theatre Arts Books.

———. 1928b. *My Life*. New York: Liveright.

———. 1981. *Isadora Speaks*. Ed. Franklin Rosemont. San Francisco, CA: City Lights.

Foster, Susan, 1986. *Reading Dancing*. Berkeley, CA: University of California Press.

———. Ed. 1997. *Choreographing History*. Bloomington, IN: Indiana University Press.

Foulkes, Julia. 2002. *Modern Bodies: Dance and American Modernism from Martha Graham to Alvin Ailey*. Chapel Hill, NC: University of North Carolina Press.

Four Pioneers. JVC Anthology of World Music and Dance. Videotape, Dance Collection, New York Public Library.

Fraleigh, Sondra Horton. 1987. *Dance and the Lived Body: A Descriptive Aesthetics.* Pittsburgh, PA: University of Pittsburgh Press.

———. 2004. *Dancing Identity: Metaphysics in Motion.* Pittsburgh, PA: University of Pittsburgh Press.

Fraleigh, Sondra Horton and Penelope Hanstein, eds. 1999. *Researching Dance: Evolving Modes of Inquiry.* Pittsburgh, PA: University of Pittsburgh Press.

Franko, Mark. 1995. *Dancing Modernism/Performing Politics.* Bloomington, IN: Indiana University Press.

———. 2002. *The Work of Dance: Labor, Movement, and Identity in the 1930s.* Middletown, CT: Wesleyan University Press.

Garaudy, Roger. 1972. *Danser Sa Vie.* Paris: Editions du Seuil.

Graff, Ellen. 1997. *Stepping Left: Dance and Politics in New York City, 1928–1942.* Durham, NC: Duke University Press.

Graham, Martha. 1923. Unpublished Letter to Ted Shawn, on display at Jacob's Pillow in August 1994. See also: Clippings files, Dance Collection of the New York Public Library on *Martha Graham* and *Clytemnestra.*

———. 1932. "The Dance in America," *Trend: A Quarterly of the Seven Arts.* 1, 1, March–April–May, pp. 5–7.

———. 1935. "Graham," in *The Modern Dance.* Ed. Virginia Stewart, 1935 and Merle Armitage, 1970. New York.

———. 1939. "The Future of the Dance," *Dance.* April, p. 9.

———. 1941. "A Modern Dancer's Primer for Action," in Rogers (1941), pp. 178–87.

———. 1950. "Martha Graham is Interviewed by Pierre Tugal," *Dancing Times.* October, pp. 21–2.

———. 1952. "The Medium of Dance." Lecture. Audiotape, Dance Collection, New York Public Library.

———. 1953. "All Dance is Contemporary and there are only Two Kinds—Good and Bad . . ." *Musical America*, February, pp. 6, 152.

———. 1958. "A Dancer's World," (Script for film) *Dance Magazine*, January, p. 5.

———. 1970. "The Background and Motivation of a Dance Artist—Martha Graham, 'How I Became a Dancer'," in Nadel and Nadel.

———. 1973. Interview with David Raher. Audiotape, Dance Collection, New York Public Library.

———. 1973. *The Notebooks of Martha Graham.* New York: Harcourt, Brace and Jovanovich.

———. 1979. "Message to Senate Appropriations Subcommittee on the NEA Appropriations, March 5, 1979," *Souvenir Program.*

———. 1984. *Martha Graham: Three Contemporary Classics* (videotape).

———. 1985. "Martha Graham Reflects on Her Art And a Life in Dance," *New York Times*, March 31, p. C1.

———. 1989. "Martha Graham on her 90th Birthday," MacNeil/Lehrer News Hour. May.

———. 1991. *Blood Memory.* NY: Doubleday.

Hanna, Judith Lynne. 1987 [1979]. *To Dance is Human.* Chicago: University of Chicago Press.

Hanna, Judith Lynne. 1988. *Dance, Sex, and Gender: Signs of Identity, Dominance, Defiance, and Desire.* Chicago: University of Chicago Press.

Helpern, Alice. 1994. *The Technique of Martha Graham.* Dobbs Ferry, NY: Morgan & Morgan.

———, Ed. 1999. *Martha Graham. Choreography and Dance: An Iinternational Journal.* Volume 5, Part 2. May.

Horosko, Marian, Ed. 1991. *Martha Graham: The Evolution of her Dance Theory and Training.* Chicago, IL: A Cappella Books.

Horst, Louis. 1961. *Modern Dance Forms in Relation to the Other Modern Arts.* San Francisco: Impulse Publications.

Irma Duncan Collection of Isadora Duncan Materials, Dance Collection, New York Public Library. [The blue notebook, ca. 1900–1903], unpublished.

Jowitt, Deborah. 1988. *Time and the Dancing Image.* Berkeley, CA: University of California.

Kealiinohomoku, Joann. 1969–1970. "An Anthropologist Looks at Ballet as a Form of Ethnic Dance," *Impulse: Extensions of Dance,* pp. 24–33.

Kendall, Elizabeth. 1979. *Where She Danced: The Birth of American Art-Dance.* Berkeley, CA: University of California.

Kisselgoff, Anna. 1991. "Martha Graham, 96, Revolutionary in the Dance World, Dies," *New York Times,* April 2.

———. 1994. "The Century of Martha Graham," *Martha Graham Centennial Program.* New York.

———. 2005. "Thoughts on the Once and Future Dance Boom," *New York Times* January 6.

Kleinman, Seymour, Ed. 1980. *Sexuality and the Dance.* Reston, VA: American Alliance for Health, Physical Education, Recreation, and Dance.

Leatherman, Le Roy. 1966. *Martha Graham: Portrait of the Lady as an Artist.* New York: Knopf.

Lloyd, Margaret. 1949. *The Borzoi Book of Modern Dance.* New York.

Loewenthal, Lillian. 1979–1980. "Isadora Duncan in the Netherlands," *Dance Chronicle,* 3(3): 227–53.

———. 1993. *The Search for Isadora: The Legends and Legacy of Isadora Duncan.* Pennington, NJ: Princeton Book Company Publishers.

Magriel, Paul, Ed. 1978 [1948]. *Chronicles of the American Dance: From the Shakers to Martha Graham.* New York: de Capo Press.

Manning, Susan. 1993. *Ecstasy and the Demon: Feminism and Nationalism in the Dances of Mary Wigman.* Berkeley, CA: University of California Press.

———. 1996. "American Document and American Minstrelsy" in *Moving Words: Re-Writing Dance.* Ed. G. Morris London: Routledge.

Martin, John. 1947. "Isadora Duncan and Basic Dance" in *Isadora Duncan.* Ed. Paul Magriel. New York: Henry Holt & Company.

Mazo, Joseph. 1977 *Prime Movers: The Makers of Modern Dance in America.* New York: Morrow.

———. 1991. "Martha Remembered: Interviews by Joseph H. Mazo," *Dance Magazine. Special Issue: Martha Graham,* July.

McDonaugh, Don. 1973. *Martha Graham: A Biography.* New York: Popular Library.

Morgan, Barbara. 1941. *Martha Graham: Sixteen Dances.* Dobbs Ferry, NY: Morgan and Morgan.

Nadel, Myron Howard and Constance Gwen Nadel, eds. 1970. *The Dance Experience: Readings in Dance Appreciation.* New York: Praeger Publishers.

Padgette, Paul, Ed. 1974. *The Dance Writings of Carl van Vechten.* New York: Dance Horizons, 1974.

Porterfield, Amanda. 1980. *Feminine Spirituality in America: From Sarah Edwards to Martha Graham.* Philadelphia: Temple University Press.

Rogers, Frederick R., Ed. *Dance: A Basic Educational Technique.* New York: Macmillan, 1941.

Roseman, Janet Lynn. 2004. *Dance was Her Religion: The Spiritual Choreography of Isadora Duncan, Ruth St. Denis and Martha Graham.* Prescott, AZ: Hohm Press.

Seroff, Victor. 1971. *The Real Isadora.* New York: The Dial Press.

Sheets-Johnstone, Maxine, Ed. *Illuminating Dance: Philosophical Explorations.* Cranbury, NJ: Associated University Press, 1984.

Shelton, Suzanne. 1981. *Divine Dancer: A Biography of Ruth St. Denis.* Garden City, NY: Doubleday and Company.

Siegel, Marcia. *The Shapes of Change: Images of American Dance.* Boston: Houghton Mifflin, 1979.

Smith-Rosenberg, Carroll. 1985. *Disorderly Conduct: Visions of Gender in Victorian America.* New York: Alfred A. Knopf.

Soares, Janet Mansfield. 1992. *Louis Horst: Musician in a Dancer's World.* Durham, NC: Duke University Press.

Sorrell, Walter. 1951. "Two Rebels, Two Giants: Isadora and Martha" in *The Dance Has Many Faces.* Ed. Walter Sorrell. Cleveland.

———. 1981. *Dance in its Time.* New York, NY: Columbia University Press.

Sparshott, Francis. 1988. *Off the Ground: First Steps to a Philosophical Consideration of the Dance.* Princeton: Princeton University Press.

St. Denis, Ruth. 1916. *Detroit Free Press.* Interview by Lucy Jeanne Price. March, 13.

———. 1932. *Lotus Light.* Cambridge, MA: The Riverside Press.

———. 1939. *Ruth St. Denis, An Unfinished Life.* New York: Harper & Brothers Publishers.

———. 1950–1959. "Seeds of a New Order," Clippings file, Dance Collection, New York Public Library.

———. 1978. "Credo," in *Dance News.* January. Clippings file, Dance Collection, New York Public Library.

Stebbins, Genevieve. 1902 [1885]. *Delsarte System of Expression.* New York: Edgar S. Werner Publishing and Supply Co.

Stewart, Virginia and Merle Armitage, eds. 1970 [1935]. *The Modern Dance.* New York.

Stodelle, Ernestine. 1984. *Deep Song: The Dance Story of Martha Graham.* New York: Schirmer Books.

Taylor, Paul. *Private Domain.* San Francisco: North Point Press, 1988.

Terry, Walter. 1947. "Ruth St. Denis, Seventy, Heads Church of the Divine Dance," *New York Times,* 7–13.

Terry, Walter. 1963. *Isadora Duncan: Her Life, Her Art, Her Legacy.* New York: Dodd, Mead & Company.

———. 1975. *Frontiers of the Dance: The Life of Martha Graham.* New York: Crowell.

The Early Years: American Modern Dance 1900–1930s. A Conference at SUNY College, April 9–12, 1981. Videotaped proceedings, Dance Collection, New York Public Library.

Trowbridge, Charlotte. 1945. *Dance Drawings of Martha Graham.* New York: Dance Observer.

Villiard, Dorothy Bird. "Martha Graham and Early Technique," Lecture at Adelphi University, Dance Collection, New York Public Library.

Wagner, Ann. 1997. *Adversaries of Dance: From the Puritans to the Present.* Urbana and Chicago: University of Illinois.

Ware, Susan. 1988. *American Women in the 1930s: Holding Their Own.* Boston: Twayne Publishers.

Index